Wine, Wealth, and the State in
Late Antique Egypt

NEW TEXTS FROM ANCIENT CULTURES

Edited by

Traianos Gagos†
James G. Keenan, Loyola University of Chicago
Terry Wilfong, University of Michigan

Settling a Dispute: Toward a Legal Anthropology of Late Antique Egypt
by Traianos Gagos and Peter van Minnen

Women of Jeme: Lives in a Coptic Town in Late Antique Egypt
by T. G. Wilfong

Wine, Wealth, and the State in Late Antique Egypt: The House of Apion at Oxyrhynchus
by T. M. Hickey

Wine, Wealth, and the State in Late Antique Egypt

The House of Apion at Oxyrhynchus

T. M. Hickey

The University of Michigan Press / Ann Arbor

Published in the United States of America by
The University of Michigan Press
Manufactured in the United States of America
⊗ Printed on acid-free paper

2015 2014 2013 2012 4 3 2 1

A CIP catalog record for this book is available from the British Library.

Library of Congress Cataloging-in-Publication Data

Hickey, Todd Michael.
Wine, wealth, and the state in late antique Egypt : the house of Apion at Oxyrhynchus / T.M. Hickey.
p. cm. — (New texts from ancient cultures)
Significantly revised version of author's 2001 University of Chicago doctoral dissertation entitled A public "house" but closed : "fiscal participation" and economic decision making on the Oxyrhynchite estate of the Flavii Apiones.
Includes bibliographical references and indexes.
ISBN 978-0-472-11812-0 (cloth : acid-free paper) — ISBN 978-0-472-02822-1 (e-book)
1. Wine and wine making—Egypt—Oxyrhynchite Nome—History. 2. Viticulture—Egypt—Oxyrhynchite Nome—History. 3. Vineyards—Egypt—Oxyrhynchite Nome—History. 4. Apion I, fl. 472–525—Family. 5. Manors—Egypt—Oxyrhynchite Nome—History. 6. Wealth—Political aspects—Egypt—Oxyrhynchite Nome—History. 7. Oxyrhynchite Nome (Egypt)—Economic conditions. 8. Egypt—Economic conditions—332 B.C.–40 A.D. 9. Egypt—Politics and government—30 B.C.–640 A.D. 10. Manuscripts (Papyri) I. Hickey, Todd Michael. Public "house" but closed. II. Title.
TP559.E3H42 2012
641.2'20962—dc23 2012000976

In memoriam T. G.

Ποῦ ἀπεσύρθηκε, ποῦ ἐχάθηκε ὁ Σοφός;

Preface

This is a significantly revised version of my 2001 University of Chicago doctoral dissertation "A Public 'House' but Closed: 'Fiscal Participation' and Economic Decision Making on the Oxyrhynchite Estate of the Flavii Apiones." The decade's delay between thesis and book is in the main attributable to external causes, but my own reservations had some influence as well. This reluctance stemmed partly from the belief that the dissertation was already easily accessible to the scholarly community through various channels—that is, that it was already "published." But it was also due to the fact that my research interests began to shift to earlier periods when I took up my post at Berkeley. Only the explosion of new publications concerning the Apiones, the appearance of additional data, and, more critically, the emergence and then ascendance of "proto-capitalist" readings of the evidence spurred my desire for revision. These novel interpretations did not always reconcile with the evidence as I had come to understand it after years of close study. Moreover, there seemed to be a reticence in this new scholarship to cite my work, presumably because it was still in dissertation form.

In the course of this project, I have accrued numerous debts. Nikolaos Gonis kindly made his voluminous scholarship on the late antique Oxyrhynchite available to me before publication and frequently shared his insights. Bernhard Palme, Giuseppina Azzarello, and Amin Benaissa were similarly generous with their unpublished research, as were Federico Morelli, Roberta Mazza, and Giovanni Ruffini. Guido Bastianini, Isabella Andorlini, Dennis Kehoe, Simona Russo, and Chrysi Kotsifou answered my questions and directed me to overlooked evidence. Comments from and discussions with

Roger Bagnall and Dominic Rathbone improved the manuscript significantly, as did the reports from the press's two anonymous readers. I am especially grateful to James Keenan, who read the entire text—and whose keen eye saved me from several slips. Any shortcomings that remain should be laid at my doorstep.

I did most of my writing at Berkeley, and I thank my supervisors and chairs—Donald Mastronarde, Charles Faulhaber, Leslie Kurke, Robert Knapp, and John Ferrari—for their guidance and support and my colleagues and students for fostering such a stimulating environment for scholarship. Brendan Haug and Alex Ladenheim provided research assistance, and Darin Jensen, our campus cartographer, patiently produced the attractive maps. Many of my research expenses were defrayed by grants from the Faculty Senate's Committee on Research. I was fortunate to be able to draft the third chapter and parts of the fourth at the Katholieke Universiteit Leuven and the Abdij Keizersberg. I thank Willy Clarysse for his kind invitation to Vlaams-Brabant and his hospitality, a warm reception that was matched by Mark Depauw, Heinz Hauben, and their colleagues and students, as well as by Paul Heilporn. At Keizersberg, Fr. Dirk Hanssens, OSB, and his *fratres* welcomed me into their community, the ideal setting for quiet reflection about the problems of the late antique Oxyrhynchite. For her sacrifices during this absence—indeed, for her extraordinary support throughout my career—I remain grateful to my wife Mary.

Working with Ellen Bauerle and Mary Hashman at the University of Michigan Press has been a joy; I wish only that I had approached Ellen sooner. This, like so many other good ideas, was the brainchild of Traianos Gagos. The sudden passing of this dear friend has left me deeply saddened—and this book without the benefit of his unselfish intelligence and wisdom. I offer it as an imperfect tribute to his memory.

Contents

Abbreviations

Greek and Coptic papyri and ostraka (as well as papyrological *instrumenta*) are referenced herein by the standard abbreviations employed in the discipline of papyrology. A current list of these may be found online at the following address (functional as of 16 August 2011): http://scriptorium.lib.duke.edu/papyrus/texts/clist.html. The most recent paper version of this document is John F. Oates et al., eds., *Checklist of Editions of Greek, Latin, Demotic, and Coptic Papyri, Ostraca, and Tablets,* 5th ed. (Oakville, CT, 2001).

Journal abbreviations conform to those of *L'Année philologique.* For ancient authors, I have employed the abbreviations of Alexander P. Kazhdan et al., eds., *Oxford Dictionary of Byzantium,* 3 vols. (New York, 1991); the few additions are unambiguous. When I have consulted an edition other than that cited in the *ODB,* this is indicated in parentheses: for example, Malal. (Thurn).

Nonbibliographic abbreviations have been avoided, and as a rule, words are not truncated until they first appear in full. Here is a list of these abbreviations:

ar. = *aroura/-ai*
Ars. = Arsinoïte or Arsinoë
art. = *artabē/-ai*
canc. = *cancellus*
choin. = *choinix/-ikes*
con. = *contubernium* or *contubernales*
d = *diploun/-a*
den. = *denarii*
ker. = *keration/-a*

knid. = *knidion/-a*
myr. = myriads
Oxy. = Oxyrhynchite or Oxyrhynchus
sext. = *sextarius/-i*
sol. = *solidus/-i*

Aids for the Reader

1. The Egyptian year consisted of twelve months of thirty days plus five (or, in leap years, six) intercalary ("epagomenal") days. Note the following (non-leap year) correspondences between Egyptian and Julian dates:

Thoth 1 = 29 August	Phaophi 1 = 28 September
Hathyr 1 = 28 October	Choiak 1 = 27 November
Tybi 1 = 27 December	Mecheir 1 = 26 January
Phamenoth 1 = 25 February	Pharmouthi 1 = 27 March
Pachon 1 = 26 April	Pauni 1 = 26 May
Epeiph 1 = 25 June	Mesore 1 = 25 July
epagomenal days: 24–28 August	

Throughout this book, reference is made to a chronological unit called the indiction. In the Oxyrhynchite nome, this period (one year long) began on Thoth 1 for chronological reckonings and Pachon 1 ≈ 1 May for fiscal reckonings. Each indiction formed part of a cycle consisting of fifteen such units (numbered 1–15); a fifteenth indiction was followed by a first. For further discussion of chronological reckoning in late antique Egypt (a complicated subject), the reader is directed to Roger S. Bagnall and Klaas A. Worp, *Chronological Systems of Byzantine Egypt*, 2nd ed. (Leiden, 2004) = *CSBE*^{2}.

2. Key measures:

1 *aroura* (the unit of land measurement) = 2,756 square meters

1 *artabē* = 40 *choinikes*. The standard *artabē* (dry measure for, e.g., grains) weighed a bit more than 30 kilograms. In the Apion dossier, *metron* and *cancellus artabai* appear most frequently; the latter *artabē* was obtained by adding a 15 percent surcharge (collected for the estate's benefit) to the former.

For liquid measures, see appendix C.

3. Money terms:

solidus/nomisma(tion): The gold coinage of the period under consideration, generally struck at ca. 4.45 grams.

keration ("carat"): A division of the solidus; not a coin but a unit of reckoning or value.

myriad: 10,000 *denarii*. A unit of account with a fixed relationship to the circulating base coinage (the *follis* and related coins). In the Apion dossier, *solidi* of various kinds are tariffed at between ca. 3,000 and 4,800 myriads.

rhopē: An assessment on a payment (probably an agio) of .5 *keration* per *solidus*.

zygon: In the papyri of the late antique Oxyrhynchite, this usually refers to a gold "standard." The interpretation of such standards is a vexed issue; Maresch 1994 and Banaji 1998 provide explanations based on accounting practice and regional metrology, respectively.

4. Specialist terminology is, as a rule, defined at its first appearance in the text. The following words are exceptions.

clarissimus, plural *-i* (Greek *lamprotatos*): Once *the* honorific designating senatorial rank, it was much devalued (and confined to the lowest tier of senators) already in the late fourth century.

gloriosissimus (Greek *endoxotatos*): With the inevitable expansion of the *illustres* (q.v.), it came to distinguish senators of the top rank within that tier. Since most honorifics were bestowed as a result of office, these were typically men active at the highest levels of imperial administration.

hyperphyestatos (Latin *excellentissimus*): An honorific marking an individual as one of the highest ranking members of the senatorial aristocracy, it is typical of the Apiones from the sixth century onward.

illustris, plural *-es*: From the late fourth century, it designates membership in the highest tier of the senatorial order, above the *spectabiles* (the middle tier; Greek *peribleptoi*) and *clarissimi* (q.v.).

magnificentissimus (Greek *megaloprepestatos*): Designating a higher rank within the *illustres* (q.v.), its status degraded during the sixth century.
pagarch: The official(s) responsible for tax collection within a nome (district).
paneuphēmos (Latin *famosissimus*): Another honorific (as *hyperphyestatos*) indicating very high rank within the late senatorial aristocracy.
patricius: A dignity that the emperor conferred on leading *illustres* (q.v.).

5. Quotations of papyri: When the Greek or Coptic text of a papyrus is uncontroversial, I reproduce it without Leiden system punctuation. Nonspecialists may "decode" the Leiden system here: http://www.lib.umich.edu/papyrus-collection/select-bibliography-papyrology#sec-Conventions (16 August 2011).

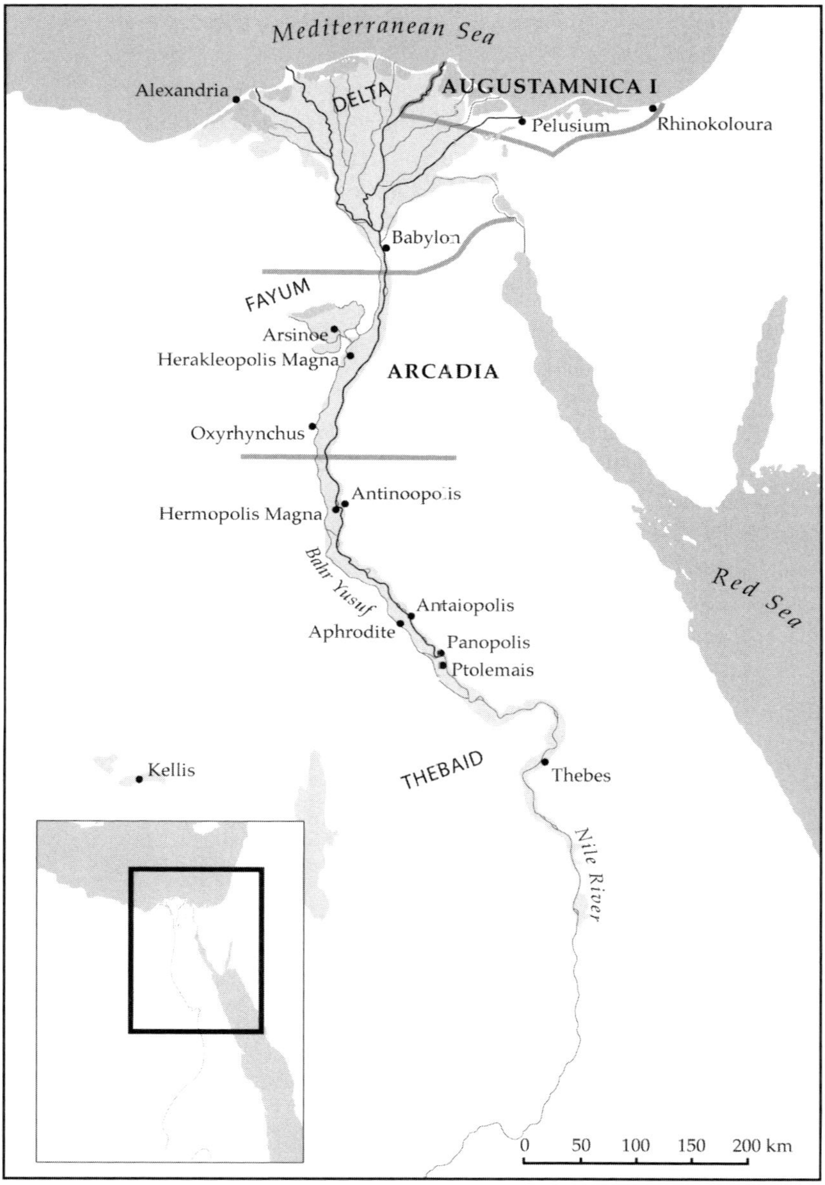

Fig. 1. Egypt in late antiquity. (Cartographer: Darin Jensen.)

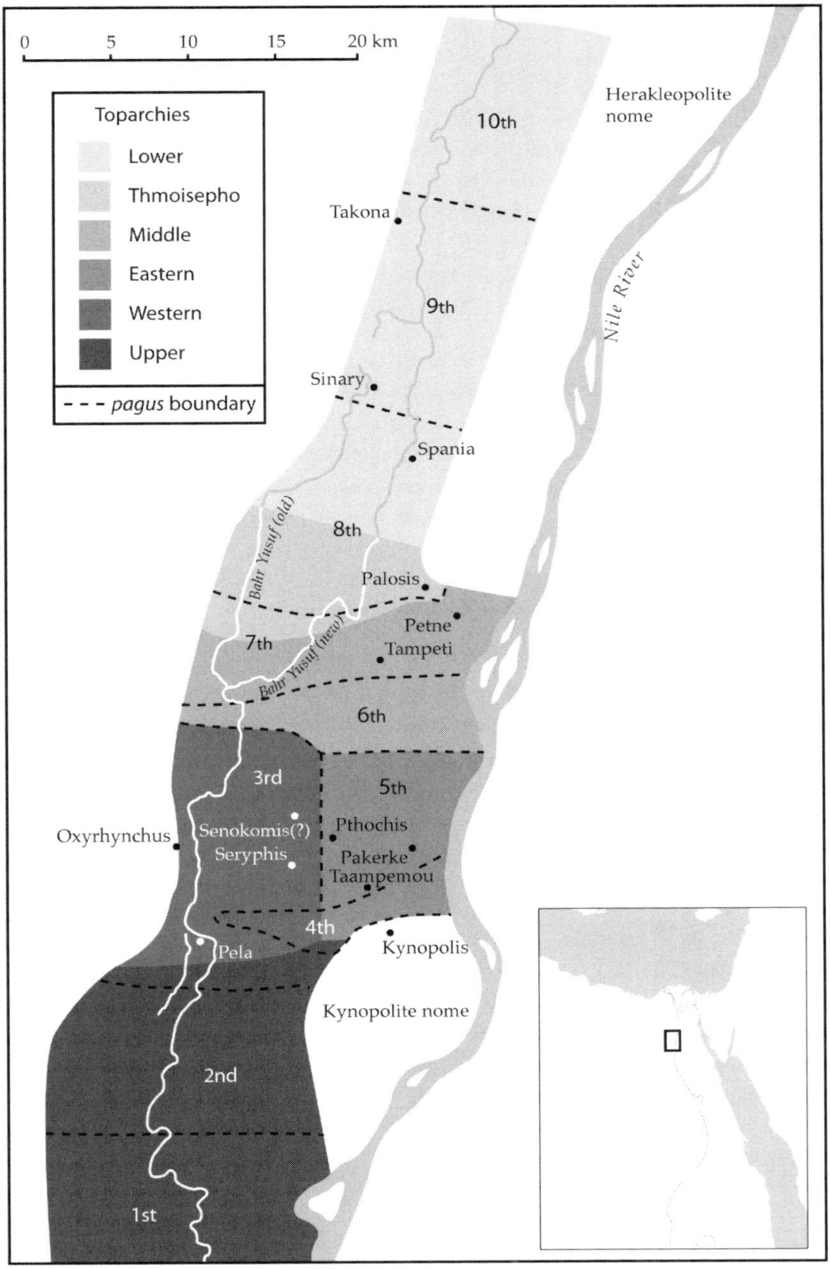

Fig. 2. The Oxyrhynchite nome during the Roman period. (Cartographer: Darin Jensen.)

Introduction

Rustica sic fiunt urbana, vetusta novella,
Publica privata . . .
—Geoffroi de Vinsauf, *Poetria nova*

Near the end of the second decade of the seventh century, the Sasanian general known as Shahrwaraz, the "boar of the empire," led an army into the Egyptian province of Augustamnica Prima, from Rhinokoloura to Pelusium and thence on to the fortress of Babylon, following a route well worn by invaders over the past millennium. Though this same commander would order the evacuation of Persian troops a mere decade later and, in doing so, restore the territory to the Romans, the brutalities of the occupation left indelible marks on the collective memory of Egyptian Christians, scars that are manifest to any reader of the extant narrative sources. Invisible in the same texts, by contrast, is evidence of the profound impact that this brief Sasanid interlude had on Egypt's agrarian economy: as the soldiers of Shahrwaraz sacked the monasteries of Pelusium, they had also begun to unravel the rural structures of late antique Egypt. These processes, once triggered, proved to be irreversible. Byzantine victory might result in the restoration of holy relics, but for the aristocrats of Egypt, there would be no *reconquista;* the Sasanid invasion marked the beginning of their end, an end that would come to full term in the eighth century, under a different (and more enduring) conqueror from the East.

The rural structures that began to disintegrate in the first quarter of the seventh century had their origins in the revolutions of the fourth, during which a new elite arose, a class that owed its development to imperial patronage, to the opportunities for enrichment and advancement afforded by the military and imperial civil service. In the course of the fifth century, a further transformation commenced: we begin to witness certain members of this elite rising through the imperial ranks and accumulating increasing amounts of land. The most successful members of this new landed aristocracy were the Flavii Apiones, a family with origins in Middle Egypt; at their apogee, the Apiones moved in the stratosphere of late antique society, at the side of the emperor himself, and owned, by some reckonings, tens of thousands of acres of agricultural land in *just* their ancestral Oxyrhynchite nome. This close association with imperial power, however, seems to have had a price: the family and then its estates disappear from the sources after the arrival of the Persians in Egypt. Yet the historical record before the invasion is rich, and the family estate at Oxyrhynchus, by virtue of hundreds of papyrus documents recovered from rubbish mounds at that site, is by far the best documented aristocratic estate from the early Byzantine world.[1]

The present monograph is devoted to this estate or, more precisely, to its production of wine. The focus on grapes may seem curiously narrow, but as we shall see, viticulture is the only agricultural "sector" through which the estate could have engaged with the market in a manner that might be considered significant from the perspective of economic history. For one goal of this study is the integration of a rich and well-contextualized corpus of papyrological evidence into scholarly discourse about the nature of the "ancient" economy.[2] Or, to put it more directly: to demonstrate how the Apion estate was managed with respect to the market, how it deployed its material and human resources and to what end. Those already familiar with the scholarly literature concerning late antique Egypt may well wonder why such an effort is necessary. It is indeed true that interest in the Apiones has surged dramatically during the

1. Mazza 2001, 20–40, lists 273 papyri that either derive from the estate archive or concern the family or its interests, but this figure has been rendered obsolete by the continued flow of editions. Perhaps half of the dossier lies unedited, and many of the published texts, especially among the accounts, have hardly been exploited.

Throughout this volume, my terminology for assemblages of ancient texts follows Martin 1994, especially 570: "I have advocated a rather broad usage of the word 'archives.' Some boundaries must be established, however, so that the term does not lose its distinct meaning. We could establish a minimal requirement: the pieces constituting an archival assemblage can in no way be the result of a chance conglomeration, even an ancient one, nor can they be a collection brought together in our time from distinct lots. They must have been the object of deliberate accumulation and classification since antiquity. If this double condition is not met, 'dossier' will be preferred to 'archives.'" A more flexible view is argued—with her usual precision—by Jördens (2001).

2. For this discourse, cf., e.g., Scheidel and Von Reden 2002; Manning and Morris 2005.

last fifteen years and that the Oxyrhynchite estate's economic activities have been central to arguments presented in critically acclaimed works,[3] but I would assert that the topic that I have signaled has yet to receive appropriate treatment. There are several reasons for this. In the first place, the use of papyrological evidence is fraught with difficulty. Typically, documents on papyrus are fragmentary and damaged, and their cursive handwriting, often rife with abbreviations, must be deciphered and then rendered into comprehensible Greek. Even the texts of the best editors are subject to correction—papyrology is the most volatile of the disciplines concerned with antiquity—and every edition must be controlled against the original manuscript. If using this material presents a challenge for specialists, what sorts of difficulties might the outsider encounter? In preparing this book, I have assumed nothing about published texts, and to the extent that access has been permitted, I have worked either autoptically or from facsimiles (photographs or digital images) of the papyri. The fruits that this "microscopic" approach has borne will be evident in the chapters that follow.[4]

Interpretation presents further difficulties. The ancient authors of papyrus texts rarely do favors for historians (they are writing for recipients or users who are "in the know"), and not all editors make a serious effort to explore these hidden contexts. In some cases, the possibilities for the reconstruction of meaning are indeed limited, but prospects are better in large dossiers like that of the Apiones. When much connectivity exists for a text, it is critical that it be interpreted *within* its corpus; that is, it must be read in such a way that it does not obliterate the social traces proffered by the other fragments of the archive. Large assemblages of papyrological texts thus present a significant challenge: their rich fragmentation must be unified, and controlling such a menagerie of text—the ancient alongside its modern interpreters—is a formidable task.[5] Failure

3. E.g., Sarris 2006; Banaji 2007.

4. One essential volume of texts pertaining (for the most part) to the Apiones, *P.Oxy.* XVI (1924), was not produced under ideal conditions; time, distance, and illness made it less reliable than other collaborations of the papyrological pioneers Grenfell and Hunt. Grenfell had been diagnosed (and institutionalized) with paranoid schizophrenia in 1920 (Montserrat 2007, 35). Hunt notes in XVI's preface, "The texts of the papyri at Cairo, most of which we had copied in the winter of 1897–8, were revised by Dr. Grenfell in 1920, when he also transcribed some of the accounts for the first time. In the preparation of the commentary the remoteness of the originals has often been a stumbling-block, since a number of textual modifications suggested themselves (especially in the accounts) which could not be verified." Cf., e.g., British Library MS Add. 59512 fol. 109 (letter from Hunt to XVI's third editor H. I. Bell, 31 December 1922). Hickey 1998 contains a batch of corrections to these Cairo texts based on autopsy, and additional installments are forthcoming.

5. It comes as no surprise, then, that papyrology is in the vanguard of Classical Studies when it comes to the application of technology.

in this imperative has been the most significant limitation of current research concerning the Apiones; texts have been read selectively, even haphazardly, without regard for their contexts. To some extent, these shortcomings must be the result of difficulties inherent in the organization of papyrological literature *sensu lato*, to the field's learning curve, but scholarly *Vorurteile* have no doubt also played a role.[6]

Of course, the interpretative unity of an archive is put in jeopardy every time a new text from it is published; this is one of the pitfalls (or pleasures, depending on one's perspective) of working with papyri. The present monograph in fact includes evidence from new (i.e., not yet published) texts that are relevant for the topic of estate wine production and the issues that are associated with it. Those addressing such matters before me have obviously been unable to make use of this information.[7] My own work, in turn, will doubtless spur correctives in the form of new texts and publications, all the more so in light of how much of the Apion dossier lies unedited. But this is much desired: I have envisioned this volume as a prolegomenon of sorts and have aimed to contribute a firm, source-based foundation for future study of the Apiones and social relations in the late antique Oxyrhynchite more generally. Accordingly, I have not shied away from providing details, and I have included substantial appendixes.

Still, this monograph differs from its antecedents in more than the quality and quantity of its evidence; its conclusions vary rather significantly. Generally speaking, recent work on the Apion estate has emphasized its modernity: its "commodification,"[8] its monetization, its rationality.[9] Some arguments of this nature have ideology at their root,[10] but the scholarship has also reacted to the concepts of *Oiken-* and *Naturalwirtschaft* (Rodbertus, Bücher, et al.),[11]

6. In addition, a false sense of the continuities of the evidence has perhaps led scholars to think that it can be sampled. Cf. below.

7. Some of it, however, was available in the dissertation on which this book is based (Hickey 2001).

8. Cf. Sarris 2006, 90, who considers the commodification of labor one of the two "most fundamental" characteristics of the Apion estate as illuminated in papyrological sources.

9. Cf. Banaji 1999, 204–6, who writes of the "implicit rationalism" in the organization and operation of the large estates of Byzantine Egypt, a "rationalism" reflected, e.g., in their substantial investments in perennial irrigation, as well as in the care with which producers structured these investments on account of the "stimulus of an expanding wine industry."

10. Both Banaji and Sarris write in the tradition of historical materialism.

11. Concepts accepted for late antiquity by both the arch "primitivist" (cf. Finley 1999, 161) and the prototypical "modernizer" (cf. Rostovtzeff 1930, 201). Against such, cf. Morelli 1996, 179: "The elements of natural economy present in Byzantine Egypt seem to derive from the occurrence of economic forms that persist to varying degrees from the pharaonic age, rather than from a regression in the use of money in the Byzantine period that led to a closed economy and reduced it to exchanges

as well as the "primitivism" of Finley and his adherents.[12] Highly influential for this revisionism has been Dominic Rathbone's rigorous, *au ras du sol* study of the third-century Appianus estate, with its diligent managers striving to control costs and achieve economies of scale and its high level of market integration;[13] indeed, Rathbone's volume was an inspiration for the present work.[14] Yet it is not clear how *rara* the *avis* Appianus was,[15] and my own reading of the Apion dossier illuminates owners and managers who, although possessing the means and many of the circumstances (structures and conditions) to reproduce Appianic "rationalism," pursued different objectives. In short, while this book does not have a stake in the debate between primitivism and modernism—the sterility of this has already been noted[16]—I do advance a position, *regarding this particular branch of the Apion estate and the extant evidence concerning it,* that others will place on the primitivist end of the spectrum.[17] For my own part, I am skeptical of the employment of Western capitalism as a touchstone for any ancient economic entity,[18] and, as a result, I have adopted an approach that is generally more descriptive than comparative.[19]

The other goal of this monograph is to provide some clarity to the relationship between the state and aristocrats like the Apiones. That the growth of substantial estates in Byzantine Egypt was facilitated by the connection of elites to the state through the bureaucracy is uncontroversial, but there has been contention about the nature of the relationship between these empowered aristocrats and the polity. The traditional—that is, the confrontational or "centrifugal"—interpretation has received its most compelling challenge from

---

in kind within self-sufficient economic units." Regardless of the merit of this argument, any claims concerning the "eternality" of Egyptian practice should give pause; cf. L. Abu-Lughod 1991, 154. Mazza 2001, 163, echoes Morelli.

12. Cf. Sarris 2006, 144–45. Banaji (1999, 213) provides a more nuanced critique: "Agriculture was history's first theatre of capitalism but because our notions have been irreducibly shaped by modern large-scale industry and the profound analysis that Marx developed in *Capital,* we only seem to grasp the history of agrarian capitalism through a sort of palimpsest. The whole debate between the 'primitivists' and the 'modernists' is essentially a misunderstanding caused by this fact, for what primitivists clearly do is measure economic behaviour by the revised edition of capitalism, so to speak."

13. Rathbone 1991. Andreau and Maucourant 1999 is the most extensive review of this seminal work; also note the articles (including a response by Rathbone) in *Topoi* 12–13 (2005): 259–313.

14. In particular, its simultaneous mastery of papyrological detail and engagement with broader historical questions. Kehoe 1992 was similarly paradigmatic.

15. Cf. Bagnall 1993c, 256.

16. Saller 2005.

17. So already Sarris 2006, 144, concerning Hickey 2001.

18. This renders me more sympathetic to the Polanyi-Finley position and institutionalists in economics ("old" more so than "new"). Cf. also the critique of the historiography of the ancient economy in Bang 2008.

19. Cf., e.g., Andreau 2002, 35.

Gascou.[20] Gascou's inventive rereading of known texts cast significant doubt on the established images of "feudal overlords with their great estates and high-sounding titles";[21] magnates once thought to have been "locked in a power struggle with the imperial government" and "capitalizing on the weakness of the central government" were in fact willing providers of public services, individuals who had voluntarily filled the gap left by the erosion of the town councils and the old system of liturgies.[22] These landowners were said to form a sort of college, dividing public charges among themselves on a system of shares (hence "model of fiscal shares" or "fiscal participation") and taking responsibility vis-à-vis the state for those not meriting a share of their own. Thus, according to Gascou, the tax payments documented in the Apion dossier not only were representative of their own possessions but also included the fiscal liabilities of the non- (or lesser) elites whose holdings fell within their share. Recently, Gascou's revisionism has itself come under attack by Sarris,[23] but the present work will demonstrate its fundamental viability; though there are problems of detail, the core of Gascou's thesis—the "fiscal participation"—is certainly valid for the Byzantine Oxyrhynchite and, accordingly, should be considered in any exploration of the society and economy of late antique Egypt.[24] I prove this not by focusing on the "public economies" of large

20. Bell 1917 and Hardy 1931 may be considered progenitors of the traditional ("feudal") model. For the alternative, Gascou 1985 (itself a development of ideas in, e.g., Gascou 1972) is paradigmatic. For the influence of Rémondon on Gascou, cf., e.g., the analysis of *P.Got.* 9 in Rémondon 1966. The work of Paul Lemerle also overshadows this fiscal school. Keenan 1993 provides an excellent overview of the historiography.

21. This description, as well as the ones following it, are from Keenan 1993. Keenan is characterizing the argument, not speaking in support of it.

Bachrach 1967 had already argued against the existence of feudalism in Byzantine Egypt. He tested the proposition, however, as one might expect that a medievalist of his era would have; in other words, he compared the testimony from the papyri to feudalism *sensu stricto,* i.e., to "military feudalism" (vassalage). Hardy et al. were never so precise in their use of the term. For a different (broader) view of feudalism (as a "mode of production" in which landlords live off the surpluses of dependent tenants and/or in reference to land-based, as opposed to tax-based, polities), see Wickham 2005, 60–61 with refs.

22. "Willing providers": insufficiently stressed is the fact that sanctioned authority need not have diminished opportunities for enrichment, and in the case of the Apiones, their often close ties to the emperor himself (cf. below) could have allowed for extraordinary liberties. Fikhman (in Keenan 1993, 142 n. 31) seems to forget this possibility (and relationship) when he suggests that the magnates usurped local authority from the state. "Erosion of the town councils" etc.: such was the view when Gascou was writing; cf. now Laniado 2002.

23. Sarris 2006, followed by Wickham (2005) and Mayerson (2006), the latter of whom writes that Gascou's model has been "largely discredited" (100). Sarris himself owes a debt to Fikhman (cf. 2006).

24. Concerning the relevance of Gascou beyond the Oxyrhynchite, see the nuanced remarks in the conclusion of Ruffini 2008.

estates, as Gascou himself did; indeed, Gascou's elision of the private, though understandable,[25] is the major weakness of his model.[26] Instead, I examine the private side of the "house" (*oikos*) of the Apiones, the large estate par excellence. Once again, the illumination of estate viticulture suffices for the task at hand.

This monograph has five chapters, as well as a conclusion and three appendixes that present some of the "raw materials" used in the study (and that are primarily intended for specialists). The first chapter, "Land," addresses the size of the estate and the historiographic significance of its area. I propose to recalculate this by estimating the area of the estate under vines. A description of the physical and administrative geography of the estate and a survey of the crops grown on it follow. Through this overview, it emerges that viticulture was the only sector in which the estate could have had significant direct engagement with external markets; as such, it is the critical sector for understanding the nature of the estate's economy. Chapter 2, "Vineland," concerns the identification of the estate's vineyards. Inter alia, it illuminates terms—*chōrion, meris, apotakton chōriōn,* and *dikaion*—that are central to this process and that have been poorly understood by previous commentators. (The sites associated with estate viticulture are themselves presented in appendix B.) The third chapter, "Labor in the Vineyard," delves into the vexed question of the estate's labor force. Recent scholarship has argued vigorously for the primacy of wage labor, but the evidence for this is found to be lacking. In light of the complexity of the issue, I adopt a descriptive approach. My reading of the texts concerning viticulture reveals tenants paying money rents, share tenants, and tenants delivering fixed-rate rents in kind; I suggest that wage workers figured most prominently in irrigation. Still, I am cautious not to draw too sharp a line between certain types of tenancy and wage labor. Chapter 3 also includes a discussion of the evolution of labor exploitation in the Oxyrhynchite and, based on the extant papyri, proposes a limited (fiscal) reading of that scholarly bugbear, the adscript colonate. The fourth chapter, "The Flow of Wine through the Estate," introduces the *oinocheiristēs,* or "wine steward," the estate employee most directly involved with the management of wine. Through his detailed accounts and other texts, I then trace the movement of wine into and

25. He considered the private to be derivative of the public: "I think that I can establish that the 'private' estate economy only represents a specific development of the public economy" (Gascou 1985, 12).

26. Cf. Bagnall 1993a, 160 and n. 58: "[C]learly their [the large estates'] public role was possible only if substantial private wealth remained to sustain the whole enterprise"; Gascou's "main thesis about the public character of much of the Great Houses' activities leads him to frame the matter too much in an either/or fashion."

out of the estate. The stewards' accounts furnish copious information about expenditures: for pious and charitable causes, to members of the civil administration and military forces, for internal expenses like salaries, and in connection with festivals and spectacles. Extant data for wine sales, in contrast, are quite limited and often indeterminate, and there is evidence for the purchase of wine. In chapter 5, I assess the importance of viticulture to the estate. I begin by positing a marketable surplus in wine of 25 percent. Even under this generous model—one lacking confirmation in the sources—the contribution of wine production to the *oikos*' economy would have been marginal; cash rents were clearly the predominant source of income. Significant direct (or autourgic) wine production, though emphasized by Sarris, is shown to be unsupported by the evidence. I then turn to the issue of the estate's relationship with the state. Here, through the employment of agricultural data (the size of the permanent workforce and an estimate of the area under vines), I consider the size of the estate. My result is an entity that is much smaller than many have proposed; accordingly, those data that suggest a large estate must document the *oikos* in its capacity as the state's fiscal agent. In other words, Gascou's "fiscal participation" argument is valid.

Who Were the Flavii Apiones?

Only rarely do the personae of documentary papyri also appear in literary or ecclesiastical sources,[27] and the Flavii Apiones easily furnish the most instances of this phenomenon.[28] I have already mentioned the extensive papyrological documentation concerning the family, and we also encounter its members with some frequency in sources like Procopius' *Wars*, the *Chronicle of Joshua the Stylite*, and the letters of Pope Gregory the Great. Of course, the information in the papyri and that of the narrative sources overlap only marginally.

The Apiones were clearly beneficiaries of the opportunities for advancement afforded by imperial service, but it should be remembered that their origins were curial, that they arose from the "class" of municipal councilors.[29] A Strategius is the first member of the family to appear in the historical record,

27. For post-Diocletianic examples, cf. the identification proposed in Zuckerman 1995, as well as Gascou 1998 and Gonis 2008, 206–7.

28. A bewildering amount of attention has been paid to the prosopography of the Apiones in the last decade, and it seems as if almost every new papyrus from the Byzantine Oxyrhynchite or Herakleopolite has some bearing on the matter. Accordingly, this account may be inaccurate on some points by the time that it appears.

29. Fikhman (1973, 18) significantly understates the importance of the "municipal aristocracy" as a feeder for the new eastern aristocracy; see, e.g., Wickham 2005, 156.

during the first half of the fifth century; in recent scholarship, he is identified as Strategius I.[30] Strategius is involved with the administration of the *domus divina* (imperial estates) when we first encounter him in 436,[31] but it is likely that he was an important local figure before this appointment in his role as *politeuomenos* (*curialis*).[32] With the *domus divina,* Strategius is initially attested as a *dioikētēs,*[33] an administrator of a division of the imperial holdings in the Oxyrhynchite. He was then promoted to *phrontistēs* (*curator*) of all of the *domus*' holdings in the nome;[34] this promotion has been connected to the transfer of the property of Theodosius II's sister Arcadia to his wife Eudocia ca. 444.[35] At some point before 460, probably before 457, the title *comes sacri consistorii* ("count of the sacred consistory") was bestowed on Strategius,[36] through which he obtained the high (for the time) rank of *vir spectabilis* (the second senatorial grade). Though he surely was a landowner of some means during much of his adult life,[37] he is first attested as a *geouchos* ("landholder") in 465;[38] the Apiones' "glorious house" (*endoxos oikos*), in its capacity as a fiscal agent, may have appeared just before this,[39] and at one point, Strategius' holdings may have extended beyond the Oxyrhynchite.[40] By 14 December 469, Strategius was dead; he is said to be "of glorious [*endoxou*] memory," which

30. The individual now called Strategius II is referred to as "I" in works published before the appearance of *P.Oxy.* L 3584–86 and the discovery of his earlier namesake. To further muddle matters, another curial Strategius is known from somewhat earlier in the fifth century; see *P.Oxy.* LXVIII 4685 and LXX 4780, 2n. The relationship of this individual to Strategius I is unknown.

31. *P.Köln* XI 459; the year 431 is a less likely possibility. See Azzarello, forthcoming, for discussion.

32. Regarding *politeuomenos,* see, most recently, *P.Worp.* 26. It continues to be applied to individuals with the two lowest senatorial ranks, e.g., Strategius II (cf. *P.Flor.* III 325). For the "emancipation" of eastern senators (and their families) from the *curia,* see Zuckerman 1998, 134.

33. Reading *P.Köln* XI 459.5–6 as διοικητῇ τῆς] θειοτά̣τ̣η̣ς̣ οἰ̣κίας. I owe this reconstruction of Strategius' career with the *domus divina* to a hypothesis in Azzarello, forthcoming, to which the reader is referred for the nuances of the evidence.

34. A position first attested with some security in *P.Oxy.* LXX 4780 (probably 457).

35. So Azzarello, forthcoming (assigning *P.Oxy.* L 3584 to the period before 457).

36. For this *comitiva,* see *CPR* XXIV, pp. 59–61, 68–71; Delmaire 1995, 29–45, has an overview of the *consistorium.* Title before 460: see, e.g., *P.Oxy.* L 3585 and *SB* XX 14091, both of which are presumed to predate the death of Eudocia since they mention her estates. Title before 457: 457 is the probable date of *P.Oxy.* LXX 4780.

37. Note *P.Oxy.* LXVIII 4687 (441).

38. *P.Heid.* IV 331.

39. Cf. Azzarello 2006, 210–11.

40. *P.Heid.* IV 331.4–5 reads, γεουχοῦντι ἐνταῦθα τῇ λαμπρᾷ καὶ λαμπροτάτῃ Ὀξυρυγχιτῶν πόλει ("landholder here in the splendid and most splendid city of the Oxyrhynchites"); there is no καί ("also") before ἐνταῦθα to indicate holdings elsewhere. Nick Gonis has kindly alerted me to an undated ineditum that has *kai,* suggesting either extension or inconsistent usage.

may indicate promotion to the top senatorial rank (the *illustres*) near the end of his life.[41]

Strategius I's only known child is a daughter named Isis, who inherited at least some of his property;[42] but the relationship of father and daughter to the figure identified as Apion I is unclear. It has been suggested that Isis married Apion I,[43] but her apparent independence as late as 482 gives pause.[44] It does seem quite probable, however, that Apion married into the family of Strategius I and that this alliance increased the fortunes of Strategius I's family significantly.[45] For Apion I was the product of a remarkably successful (and old) curial line, the Septimii Flaviani of the neighboring Herakleopolite;[46] his father, Flavianus I, had risen from being a member of a leading local family to the office of *comes sacrarum largitionum*.[47] (Strategius II occupied this same post two generations later, an event once considered a milestone in the rise of the Apion family.)

It appears that Apion I is first attested in 472, fulfilling the liturgy of *defensor civitatis* in Oxyrhynchus on behalf of his deceased father-in-law.[48] These early texts concerning the *defensio* are also the last papyri to be addressed to Apion directly, which may indicate his departure to Constantinople.[49] A series of Herakleopolite rent receipts from 483 reveal his membership in the second senatorial grade (*spectabilis*); he is named alongside his younger brother Flavianus (still a *vir clarissimus*), presumably because they had corporate ownership of their father's estate.[50] By 489 Apion was being called *excellentissimus*

41. *P.Oxy.* LXIII 4390. This posthumous reference may not, however, be accurate; cf. *P.Oxy.* LXVIII 4693, 3n.

42. See *P.Oxy.* LXIII 4390 (469) and 4391 (471) and *P.Thomas* 26v (undated), all of which refer to affairs on her estate.

43. A hypothesis first proposed by Gonis in *P.Oxy.* LXVII 4614, 2n., and since picked up by others (e.g., Azzarello 2010) with somewhat less caution.

44. The text in question is forthcoming in *P.Oxy.* and was kindly brought to my attention by Nick Gonis.

45. Gonis 2004b is fundamental for the life and career of Apion I.

46. See Gonis 2000b, 72–73, for the significance of the *gentilicium* Septimius.

47. Palme (2008, 155) summarizes his rise as follows: "One of the prominent members of the curial stratum made the crucial step from the municipal level into imperial service, first holding one or two governorships in Egypt (both with the rank of *praeses*), before successfully taking over a governorship with the rank of *consularis* outside his homeland. From there he advanced into Constantinopolitan posts and, finally, even into the palatine *comitivae consistoriana* and *sacrarum largitionum*." His brother Ptolemaeus was almost as accomplished.

48. So Azzarello 2010, interpreting *P.Oxy.* XVI 1886 (to which she assigns the date cited above) and the unpublished P.Wash.Univ. inv. 26.

49. Azzarello (2010, 43) suggests that *P.Bingen* 135 (483) is the terminus post quem for this; I am inclined to accept her proposal, but it should be noted that even that text involves an intermediary.

50. *P.Bingen* 135. Corporate ownership: Cf. Banaji 1999, 205.

(*hyperphyestatos*) and was possibly an honorary consul (*ex consulibus*).[51] The honorary consulate is certain by 497,[52] and Apion is identified as a *vir gloriosissimus* a few years before this date.[53] Apion then begins to appear in narrative sources: several works relate that he was involved (with logistics) in the Byzantine campaign around Amida following the fall of that city to the Sasanian king Kavād in 502;[54] the testimony of various authors (e.g., Malalas) that he was *praefectus praetorio Orientis* (no doubt *vacans*) during this campaign is now confirmed in a papyrus.[55] By 503 Apion was also a *patricius.*[56] In 510, however, the emperor Anastasius put an end to his advancement:[57] he exiled Apion, probably not as a result of his conduct during the war but for his religious sympathies, and he may have confiscated his lands.[58] Theodore *lector* records that Anastasius ordered both Apion and a son named Herakleidas to be ordained. Apion, who was to receive priestly orders from the bishop of Nicaea, is said to have reacted violently; in contrast, Herakleidas, who was to be ordained a deacon at Prusa, entered the church joyfully.[59] The emperor Justin recalled Apion and made him praetorian prefect;[60] this must have happened fairly early in Justin's reign, for Apion is attested in the position at the end of 518.[61] According to the testimony of his own son Strategius, Apion had been converted to Chalcedonianism (from miaphysitism) by Justinian and

51. *P.Flor.* III 325 + *BL* VII 53.

52. *Stud.Pal.* XX 129. Gonis (2004b, 176) suggests that it was purchased under Zeno (474–91).

53. *P.Eirene* II 12 (492). For the "evolution" of Apion's titles, see *P.Oxy.* LXVII 4615, 4–5n., where the apparent replacement of *endoxotatos* by *paneuphēmos* is attributed to regional variation. But perhaps *paneuphēmos* indicates his rise to the patriciate.

54. See Procopius, *Wars* 1.8.5 (predictably Atticized); *JoshStyl* (Trombley-Watt) 54 (pp. 65–66), 70 (p. 88); Theod. *lect.* (Hansen) 482 (p. 137); Malal. (Thurn) 16.9 (p. 326); Theoph. AM 5997 (p. 146), AM 5998 (p. 148).

55. *P.Oxy.* LXVII 4615.4 (505), in which he is styled *apo . . . eparchōn.*

56. Malal. (Thurn) 16.9 (p. 326); Procopius, *Wars* 1.8.5.

57. See MGH *AuctAnt* 11 (= Marcellinus *comes*), p. 97; Theod. *lect.* (Hansen) 482 (p. 137); and Lydus (Schamp) 17.3 (vol. 2.1, pp. 64–65).

58. Conduct: So Hardy 1931, 26 (cf. Theoph. AM 5998 [p. 148]). Apion's association with Severus of Antioch may have been to blame; cf. Mazza 2001, 56. Confiscation: Only mentioned by Lydus (see previous note).

59. Theod. *lect.* (Hansen) 482 (p. 137). Theodore ascribes Apion's reaction to heresy and pederasty. It has not yet been noted that Apion's forced ordination is also cited in Peter Damian's epistle LXXII (Reindel, vol. 2, p. 331); the Benedictine monk says (p. 330) that his source was an ecclesiastical history *quam Romanus quidam scriniarius scribit.* The identity of this work is unknown.

60. Cf. *Chron. Pasch.*, a. 519 (p. 612); Malal. (Thurn) 17.3 (p. 337). It is also noted by Peter Damian (see previous note).

61. *CJ* VII.63(3), 1.xii.518. Mazza (2004, 266), who attributes Apion's rehabilitation to "a tie of political and personal friendship that arose in the time of the military campaign against the Persians," suggests a role for the family in Justin's rise to power.

Theodora.[62] Their probable date of marriage (525) is the terminus post quem for his death, while Strategius' testimony, which occurred while he was representing the emperor Justinian at a meeting of Chalcedonian and Syrian Orthodox bishops early in 532, provides the terminus ante quem.[63]

Apion's son Herakleidas (or –ēs) is an evanescent figure. He presumably was older than Strategius II,[64] but Strategius is much better documented. Beyond Herakleidas' ordination, we know only that he was a *principalis* at Herakleopolis at some point in the period before his diaconate.[65] His brother Strategius II is first attested in 489, at which time he is styled *vir clarissimus* and *curialis,* as well as identified as an Oxyrhynchite landowner (*geouchos*).[66] On this occasion (and another), he acts through an intermediary, most likely as a result of his minority.[67] Given Apion I's virtual absence from the Oxyrhynchite documentation, it seems probable that Strategius acquired his property through his grandfather Strategius I or his mother, though the precise mechanism of acquisition is unclear.[68] By 497 Strategius was an independent actor.[69] He was now titled *magnificentissimus et gloriosissimus comes devotissimorum domesticorum;*[70] this *comitiva* made him a *vir illustris* (and eligible for the "active" senate) at, it would seem, a young age.[71] He likely was serving as *comes largitionum* of the diocese of Egypt in January of 492 or 507.[72] Between 518 and 524, Strategius was appointed *praefectus Augustalis,*[73] a position that may account for an increase in rank that is first attested in 523; at some point before this date, he had acquired the honorary titles of *magister militum* and *ex con-*

62. Strategius' words were recorded in a (Greek) letter, now lost, of Innocent of Maroneia to Thomas *presbyteros* of Thessalonica; see *ACO* IV.2, pp. 169–70, for the surviving Latin translation.

63. For the latter date, cf. Gascou 1985, 63 n. 352.

64. So Azzarello 2010, 40.

65. *CPR* VI 79.1–2: Σεπτιμίῳ Ἡρακλείδῃ Ἀπίωνος πρωτεύοντι Ἡρακλέους πόλεως, "to Septimius Herakleides, the son of Apion, *principalis* of Herakleopolis"; note also *P.Rain.Unterricht* 81.12, 14. *Principalis:* For these municipal elites, see, mostly recently, *P.Köln* XI 460, pp. 239–43.

66. *P.Flor.* III 325.

67. Cf. Gonis 2004b, 176. The other text, which attests Strategius executing the liturgy of *riparius* (which often fell on the Apiones' *oikos;* cf., e.g., Mazza 2001, 83), is *P.Oxy.* LXVII 4614 (before 497).

68. See Gonis 2004b, 176–77. His mother seems much more likely: if Strategius I had been directly responsible, Strategius II would have had to have been born before 469 (the terminus ante quem for Strategius I's death).

69. *P.Oxy.* XVI 1982.

70. He bears this title in *P.Oxy.* XVI 1982 (497), *P.Oxy.* LXVII 4615 (505), and P.Harr. inv. 550a (Oxy.; 503 or 518). It is also probable in *CPR* XIV 48 (Herakleopolite; 506). For the *comitiva,* see Palme 1998a, 104–16.

71. See, further, *P.Oxy.* LXVII 4615, 3–4n.

72. *SB* XX 14674.36, with Gascou and Worp 1988, 105. The date 522 is also possible, depending on when he became *Augustalis.*

73. The terminus post quem assumes that the appointment was made by Justin. For the terminus ante quem, see *Edict* XIII.15–16 with Rémondon 1955.

sulibus.[74] Strategius was now also in control of land outside the Oxyrhynchite,[75] holdings that no doubt included property in his paternal Herakleopolite,[76] but it is not clear if these represent lands from his father (coming to him in the "confiscation" or after Apion's rehabilitation or death) or were acquired through other means.[77] He possessed the dignity of *patricius* possibly as early as 528[78]—certainly by September 530[79]—and seems to have become *comes sacrarum largitionum* shortly after the meeting of Chalcedonian and Syrian Orthodox bishops (in 532) mentioned above.[80] While *comes*, Strategius oversaw the reconstruction of Hagia Sophia, which was rededicated on 27 December 537.[81] Like his father, he also had some involvement with external affairs; Procopius records that Justinian sent Strategius and a certain Summus ("former commander of the soldiers in Palestine") to settle the so-called Strata dispute (ca. 539) between the Lakhmid (and Sasanian client) al-Mundhir and the Ghassanid *foederatus* al-Ḥārith.[82] This assignment, as well as his earlier involvement with the bishops, no doubt indicates that he had the emperor's confidence; a later source in fact records that Strategius and Justinian had a very close relationship.[83] Strategius probably died at the beginning of 542. In the bureaus of

74. *P.Oxy.* XVI 1984; his epithets are now *endoxotatos kai hyperphyestatos.*

75. Cf., e.g., *P.Oxy.* LXVII 4616.3–4 (525): γεουχοῦντι καὶ ἐνταῦθα τῇ λαμπρᾷ Ὀξυρυγχιτῶν πόλει, "landholder also here in the splendid city of the Oxyrhynchites."

76. Herakleopolite holdings are certainly attested by September 530 (*P.Oxy.* LXX 4784.4–5: πρωτεύοντι κατά τε τὴν Ἡρακλεοπολιτῶν καὶ κατὰ ταύτην τὴν λαμπρὰν Ὀξυρυγχιτῶν πόλιν, "*principalis* in both the city of the Herakleopolites and this splendid city of the Oxyrhynchites"), possibly as early as 528 (*P.Oxy.* LXX 4783). Note, however, *CPR* XIV 48 (506), which has him undertaking the ripariate in Herakleopolis, but this may be comparable to his father's service as *defensor civitatis* in the Oxyrhynchite (on which, see above).

77. Cf. *P.Oxy.* LXVII 4616.3–4n. I suspect that they were inherited.

78. If *P.Oxy.* LXX 4783 dates to 528.

79. *P.Oxy.* LXX 4784.

80. Cf. Innocent of Maroneia in *ACO* IV.2, p. 170: *nunc comes divinarum ubique largitionum est.* According to the *Breviarium* of Liberatus (*ACO* II.5, p. 110), he was already *comes sacrarum largitionum* when the meeting began. It seems a stretch to suggest (as, e.g., Gascou 1985, 64, based on *ACO* IV.2, p. 169, and the unreliable *Patria* 1:78–79, 89, 105, 2:259–60) that Strategius was a "temporary" *magister officiorum* at this meeting; he was simply attending in the *magister*'s stead; cf. also Harvard syr. 22.3 (in Brock 1981, 92).

81. For Strategius' involvement, see *Patria* 1:84–85. Hagia Sophia had been destroyed in the Nika riots of 532; the exact date of rededication is given by Theophanes (AM 6030 [p. 217]) and "should be accepted" (so Mango-Scott, p. 316).

82. Procopius, *Wars* 2.1.1–11; Strategius is there identified as "magistrate of the royal treasures," τῶν βασιλικῶν θησαυρῶν ἄρχοντι (9). Malal. (Thurn) 18.61 (p. 390) and 68–69 (pp. 393–94) may well reference an earlier mission performed by Strategius.

83. Strategius is called, e.g., "adopted brother of the emperor," ὁ τοῦ βασιλέως ἀδελφοποιητός (*Patria* 1:79) and "spiritual brother of the emperor Justinian," πνευματικὸς ἀδελφὸς τοῦ βασιλέως Ἰουστινιανοῦ (1:85). For *adelphopoiēsis,* which figures in Boswell's controversial *Same-Sex Unions,* see Rapp 1997.

his Oxyrhynchite estate, he was thought to be alive on 18 February 542;[84] the elevation of Peter Barsymes to count of the sacred largesses suggests a decease before 1 March.[85]

At Strategius' death, his son Apion II was still under the supervision of his mother, Leontia.[86] The precise nature of her guardianship is unclear, but it seems more likely to have been a case of *tutela impuberis* than *cura minoris.*[87] Apion's birthday is known to have been Tybi 28 (23 January),[88] and since he appears in control of the Oxyrhynchite estate on 9 October 543,[89] he presumably reached *pubertas* (or possibly full majority) in January of that year. In other words, Apion was probably born in 529 (if we accept *tutela impuberis*), at the very earliest in 518 (under *cura minoris*).[90] His father was over thirty at the time of his birth.[91]

With Apion II, the family's fortunes reach their apogee. In 539, at a rather young age under any calculation, Apion obtained the empire's highest honor, the ordinary consulship. He was one of the last citizens to be granted this dignity, and we can only speculate about the politics behind it.[92] A magnificent ivory diptych of Apion's (fig. 3), still preserved in the cathedral at Oviedo, depicts his youth and further reveals that he was a *comes devotissimorum domesticorum,* like his father early in his career.[93] The extant papyri of the Oxyrhynchite estate indicate a progression in his titulature: both *paneuphēmos* (Latin *famosissimus;* before 19 October 550) and *patricius* (before 3 October 565) are added to the Greek renderings of *ex consulibus ordinariis* (or *consul ordinarius*) that typically appear.[94] Apion was also *prōtopatrikios* during his lifetime; among the honors

84. *P.Oxy.* LXIII 4396.

85. See *Edict* VI.6. It has been suggested that Strategius was a victim of the Justinianic plague (e.g., Mazza 2004, 267), but his death seems to have occurred a bit early for the outbreak in Constantinople (for the date of which see Stathakopoulos 2004, 114). In any case, it should be remembered that he could have been rather aged at his death (and was certainly over fifty).

86. For speculation regarding Leontia's family, see *P.Oxy.* LXIII 4397, 120n.

87. Cf. Beaucamp 2001, 168–70.

88. *PSI* VIII 953.80–81 (in a leap year: January 24).

89. *P.Oxy.* XVI 1985.

90. Under *venia aetatis,* Apion could have been born between 519 and 522.

91. It would seem unlikely that Apion was his first child, but the "subsidiary" lines of the family's stemma remain shrouded in mystery.

92. Beaucamp (2001) emphasizes the Apiones' personal relationships with the imperial house; Mazza (2004, 270–71) suggests that Strategius II was too shrewd to accept the dignity.

Ruffini (2008, 97–98) seems to believe that Strategius was already dead at the time of the appointment.

93. See Delbrück 1929, 1:150–51, 2: pl. 33; Volbach 1976, no. 32. The inscription is *CIL* II 2699.

94. *Paneuphēmos:* first secure attestation in *P.Oxy.* I 133. *Patricius: PSI* III 191.1–2; Apion did not have this rank on 12 March 564 (*P.Oxy.* LXX 4787). *Hyperphyestatos* appears alongside the consular references from the start.

Fig. 3. Apion II's consular diptych (539). (Courtesy of the Museo de la Iglesia, Catedral de Oviedo.)

conferred by this title, which was reserved for the eldest and most esteemed senator, was the right to speak first in the senate.[95] There is no evidence whatsoever that Apion held any offices, in Egypt or elsewhere.[96]

The bureaus on Apion's Oxyrhynchite estate considered him alive on 25 March 578,[97] but news of his death reached Middle Egypt before 19 January 579.[98] His succession was clearly a complicated matter: for eight years, the papyri from Oxyrhynchus are addressed to a collective of heirs, described first as "heirs and men" (κληρονόμοι and ἄνδρες),[99] then as "successors and men" (διάδοχοι and ἄνδρες),[100] and finally as "successors and people" (διάδοχοι and πρόσωπα).[101] Consortial ownership in the wake of a decease was not unusual; Banaji, in fact, sees it as emblematic of the eastern aristocracy of this period, the means by which these elites avoided "the fragmentation of holdings through inheritance which had drastically undermined the economic strength of the former municipal families."[102] Only after December 587 do Apion's daughter-in-law (or daughter) Praeiecta and her sons Apion and Georgios emerge in control of the Oxyrhynchite estate.[103] Presumably this signifies that Apion II's other successors had received property elsewhere (or some other form of compensation).

---

95. See Mazza 1997 with *P.Oxy.* LXX 4792, 5–6n. The title appears only in certain posthumous texts.

96. Pace, e.g., Sarris 2006, 19: "By the late 540s he is described as *patricius* and, by c. 550, *dux Thebaidis.* In 556 Apion III [*sic*] can be seen bearing the titles of *magister utriusque militiae* and pagarch of Arsinoe in the Fayum." Apion III is a better candidate for *dux* (cf. *P.Oxy.* LXX 4787, p. 99 and refs.), while the pagarchy (and the title *stratēlatēs*) belong to another Apion (cf. *P.Oxy.* LXX 4786, pp. 92–93).

97. *P.Oxy.* LXX 4791.

98. *P.Oxy.* LXX 4792.

99. 19 January 579 (*P.Oxy.* LXX 4792) through 8 October 583 (*P.Oxy.* LXX 4796).

100. 11 January 584 (*P.Oxy.* I 137) through 20 November 584 (*SB* XVI 12484).

101. 26(?) September 586 (*P.Oxy.* LXX 4798) through 29 December 587 (*P.Oxy.* XVI 1988 = *P.Oxy.* LXX, pp. 137–38). It would seem that πρόσωπα indicates the presence of women, but see *P.Oxy.* LXX 4788, 4n.

Those seeking further details concerning Apion's succession, one of Byzantine papyrology's more vexing problems, should see Palme 1998b; Beaucamp 2001, 171–78; and Mazza 2004, 273–80.

102. Banaji 2000, 96. One of the editors of this series reminds me that consortial ownership was not an elite phenomenon exclusively.

103. P.Vindob. G 13,381+22,003 indicates that Praeiecta was married to a Strategius, but which one (the name is not rare) is unclear. The chronology eliminates Strategius *paneuphēmos* (cf. Palme 1998b) from consideration. (I thank Bernhard Palme for allowing me to see the Vienna text before its publication.)

P.Oxy. XVIII 2196r, a *pronoētēs'* (steward's) account for a fifth indiction, is addressed to Praeiecta and her sons; presumably the fifth indiction in question was 586/87. The account should have been written up in the first part of the sixth indiction (587/88), but *P.Oxy.* XVI 1988 suggests that this did not happen before the end of 587. The format of the account is odd, however, and it is probably scribal practice (cf. pp. 127–28 below).

Apion III acquired sole control of the Oxyrhynchite estate before 27 March 593.[104] His brother Georgios disappears from the documentation by November 590,[105] and it has been suggested that he died.[106] His mother is last attested in Oxyrhynchite texts a little more than two months later,[107] but her *oikos* (no children are mentioned) is referenced in a Herakleopolite document dated 17 September 591.[108] Apion's own Herakleopolite *oikos* is first attested on 26 August 593.[109] The Herakleopolite text referring to Praeiecta could signify the final dissolution of the consortium of heirs into individual *oikoi,*[110] but it might also be the case that Praeiecta is present in our documents largely because Apion was unable to act independently on account of his youth.[111] In any case, Apion married well: a letter of Pope Gregory the Great reveals that Apion's wife was Eusebia, a woman from the line of the Anicii.[112] Other letters from Gregory reveal that the couple had children, including a boy named Strategius (IV);[113] in one of these, Gregory advises Eusebia to proceed with caution in the wake of the usurpation of Phocas (popular in Rome, but considerably less so in Constantinople).[114] Though the estate itself is well documented under Apion III, the papyri reveal little about his activities; here, it is sufficient to note that he had become a *patricius* by the end of 604 or the beginning of 605 (very probably under Phocas).[115] Apion III was thought to be alive as late as 5 July 619 and known to be dead on 12 January 620, close enough to the beginning of the Sasanian invasion for it to be suggested that he might have fallen fighting the Persians.[116] The Oxyrhynchite *oikos,* however, seems to

104. *P.Oxy.* XVIII 2203.

105. Georgios is not mentioned in *P.Oxy.* XVI 1989 (3.xi.590).

106. Cf. Palme 1998b, 305.

107. *P.Oxy.* XVI 1990 = *P.Oxy.* LXX, pp. 140–42 (12.i.591).

108. *P.Erl.* 67.

109. *Stud.Pal.* III^2 86.

110. So Palme in *CPR* XXIV 25, pp. 148–49. The unpublished P.Vindob. G 13,381+22,003 would seem to imply that Apion acquired further possessions via inheritance from his parents; both had died before 10 April 607.

111. It may be recalled that his grandfather Apion II was probably born in 529, so the chronology does not undermine the hypothesis. *P.Erl.* 67 could refer to Praeiecta's own property (i.e., holdings not inherited from Apion II).

112. Gregory, *Reg. ep.* II 27 (April 592); cf. Cameron 1979, 225–27. Her mother, Rusticiana, certainly had property in Sicily and doubtless possessed significant holdings elsewhere in the West.

113. Gregory, *Reg. ep.* IV 44 (August 594), VIII 22 (May 598, the first mention of Strategius), XI 26 (February 601), XIII 26 (February 603), XIII 35 (June 603). Another son may appear on a Cypriot lead seal; cf. Metcalf 2004, 210 (a reference that I owe to Chrysi Kotsifou).

114. Gregory, *Reg. ep.* XIII 35.

115. *P.Oxy.* LVIII 3941 (27.xii.604–25.i.605). *PSI* III 179 (probably 25.xii.602; cf. *P.Oxy.* LVIII 3933 2–3n., no. 10 + *P.Thomas,* pp. 258–59) provides the terminus post quem.

116. *P.Oxy.* LVIII 3959, pp. 116–17. This assumes that he was even in Egypt.

have continued to function as an economic unit until at least August 626,[117] but it does not resurface after Heraclius' *reconquista.*

The Apion Papyri as Sources: Some Specific Issues

Sarris has recently remarked, "I have found that the Apion archive is not really amenable to the sort of deep, structural analysis that has informed some of the best papyrological studies of recent years."[118] This is an excessively pessimistic assessment, but it is true that the Apion papyri have limitations as evidence and pose some formidable challenges.[119] Any consumer of scholarly literature concerning the family or its *oikos* should be aware of the origin of the papyri that underlie such studies. The bulk of this evidence derives from "wastepaper," from documents that were deliberately discarded in the rubbish heaps outside the city of Oxyrhynchus, presumably by staff of the central bureaus of the estate.[120] The papyri that we possess today survived the hazards of their environment for the next fourteen centuries more or less intact and then happened to be excavated; it should be quite clear, in other words, that we possess a mere fraction of the original material that was discarded (and an even smaller portion of the total holdings of the estate bureaus).[121] The papyri

117. *P.Oxy.* LXVIII 4704 (29.viii–27.ix.626).

118. Sarris 2006, vii.

119. Some of the *general* problems involved with the use of papyri as historical sources have already been mentioned; for further discussion of such issues, see, e.g., Bagnall 1995 and Hickey 2009.

120. These offices were perhaps located in buildings connected with the estate house (on which cf. Mazza 2001, 84–85). It is conceivable that there was a mass liquidation of records when the estate was dissolved during the Persian occupation, but I am skeptical because large quantities of papyri had value (as fuel at a minimum). For exemplary analysis of an earlier "wastepaper archive," see Verhoogt 1998, 22–49.

Some of our extant papyri—probably a good many more than has been realized—had a second life before disposal: they seem to have been reused in the Apiones' bureaus for scribal training. A modern interpreter must be aware of these later phases of use. Cf. Harrauer and Sijpesteijn 1986 and Hickey 1999, 325–26.

121. B. P. Grenfell and A. S. Hunt of Oxford found the lion's share of the papyri, and their activities at Oxyrhynchus are documented in Grenfell 1896/97 and Grenfell and Hunt 1902/3, 1903/4, 1904/5, 1905/6, and 1906/7 (all now conveniently reprinted with supplementary notes in Bowman et al. 2007, 345–68). These papyri are now dispersed between Oxford (which has the vast majority of them), Cairo, and the consortium of institutions that helped fund their acquisition. For the Italian finds at Oxyrhynchus, see Breccia 1932 and 1933 as well as Pintaudi 2007 (with refs.). The papyri unearthed during these campaigns are in Florence, at both the Biblioteca Medicea Laurenziana and the Istituto Papirologico "G. Vitelli." Flinders Petrie also excavated at Oxyrhynchus, cf. Petrie et al. 1925, 12–18 (reprinted in Bowman et al. 2007, 50–69); he is the source of the few Apion papyri in the collection of Washington University in St. Louis. Other texts (e.g., those among the Ianda papyri in Giessen) were acquired on the antiquities market.

that are currently accessible to the public have been further impacted by human agency: editors naturally prefer easy-to-decipher contracts (and other formulaic texts) to decontextualized letters, to say nothing of dry accounts, the texts most critical to economic analysis of the estate.[122]

The historian attempting synthetic study of the Apiones is glimpsing but one node, albeit the central one, of a local network; this node, in turn, connected the local network to its analogues centered in Alexandria and Constantinople.[123] Encounters with traces of the family's economic activity beyond the Oxyrhynchite are rare (and largely confined to the neighboring Kynopolite and Herakleopolite nomes), to say nothing of activity beyond the preservative sands of Upper Egypt.[124] Even within the Oxyrhynchite itself, our knowledge is largely limited to settlements within the Apion orbit, hence the traditional historiographical emphasis on the "great estate"—and the elision of the (more) independent village.[125] Moreover, day-to-day operations in the known periphery only come into view when *événements* bring documents from the hinterland to the city. Inevitably, this has skewed our perception of the content of the estate's documentary production. For example, we possess large central accounts recording the expenditure of bricks and axles for irrigation machinery (*mēchanai*),[126] as well as a plethora of receipts for replacement parts from

---

The account of Grenfell and Hunt's 1897 discovery, a virtual sine qua non for research on the Apiones, merits quotation in part.

> The third and by far the greatest find, that of the Byzantine archives, took place on March 18th and 19th, and was, I suppose, a "record" in point of quantity. On the first of these two days we came upon a mound which had a thick layer consisting almost entirely of papyrus rolls. There was room for six pairs of men and boys to be working simultaneously at this storehouse, and the difficulty was to find enough baskets in all Behneseh to contain the papyri. At the end of the day's work no less than thirty-six good sized baskets were brought in from this place, several of them stuffed with fine rolls three to ten feet long, including some of the largest Greek rolls I have ever seen. (Grenfell 1896/97, 8)

An additional twenty-five baskets were filled the following day. Grenfell later identified this as the "Apion archive"; cf. Bowman et al. 2007, 350, where it is also suggested that the mound is no. 11 on Darbishire's map (*P.Oxy.* L, p. vii).

122. Grenfell and Hunt themselves furnish an admirable exception, given the diversity of the papyri that they selected for publication.

123. The Apiones' high status suggests landholdings outside of Egypt, but there is no firm evidence for such.

124. Beyond the Oxyrhynchite: Kynopolite possessions seem to have been administered from Oxyrhynchus (cf., e.g., *P.Oxy.* XVI 2025, noting l. 13); for the Herakleopolite, cf. Mazza 2001, 38 (as well as the family history that I presented in the preceding section). Beyond Upper Egypt: Cf., e.g., *PSI* VIII 953.72–74; *P.Oxy.* XVI 1904.

125. Cf. the remarks in Ruffini 2008, 248–51.

126. *P.Oxy.* XVIII 2197, XIX 2244. For the moment, I leave aside the fiscal element to irrigation, which no doubt enhanced its documentation.

individual *geōrgoi* (literally, "farmers"), who typically had to go to the city to request this equipment.[127] In contrast, details about the exploitation of the plots that this machinery watered are limited. The machinery receipts reveal the crops grown, but we know relatively little about the daily activities of the people and animals who worked the land.[128] It is not even clear what the precise relationship between those "going up to the city" and the Apiones was: were they its tenants or its employees (or something outside the *locatio conductio* norm)? In other words, what does *hypo* signify in the receipts' τὴν ὑπ' ἐμὲ γεουχικὴν μηχανήν (literally, "the landowner's machinery under me"), and does it always mean the same thing?[129] Most of the equipment recipients are *coloni adscripticii*, and their poor lot might lead us to argue for employees.[130] When we then consider other factors, such as the general prominence of *adscripticii* in the extant Apion papyri[131] or the real dearth of lease contracts from the estate,[132] a conclusion that the *endoxos oikos* "relied quite substantially on wage labour" seems perfectly reasonable.[133] But might this putative primacy of wage labor be a fiction of survival and praxis, a distortion amplified, furthermore, by the inherent fiscality (and, of necessity, documentation) of the *adscripticii?*[134] It is certain, in any case, that whole categories of labor have gone unwritten: in particular, the extremes of permanence (e.g., customary arrangements) and transience (casual labor) tend to be linked to silence. Thus the absence of leases might be more satisfactorily explained by the use of a rent roll and *relocatio tacita.*[135]

---

127. Cf. *SB* XXII 15364.18–21, addressed to the heirs of Apion II: ἀνῆλθον [so corrected by eds.; read ἀνελθὼν] ἐπὶ τῆς πόλεως ἠξίωσα τὴν ὑμῶν ὑπερφύειαν διὰ τῶν αὐτῇ προσηκόντων ὥστε κελεῦσαι παρασχεθῆναί μοι τὸν αὐτὸν μέγαν ἐργάτην, "After I went up to the city, I requested your excellency, through its representatives, to order this same large cogwheel to be provided to me." For such traffic from countryside to city, cf., e.g., Keenan 1981. I discuss these receipts further below (pp. 50–51, 83–84).

128. Contrast Rathbone 1991, the foundation of which is documentation from a rural unit (*phrontis*) of a large estate.

129. Cf. the discussion in Gascou and MacCoull 1987, 111.

130. Cf. Banaji 1997, 96; Sirks 2008, 133. Banaji (1999, 210–11) emphasizes the receipt of cash advances (*prochreiai*), which he considers wages, by these individuals. These matters are addressed in detail in chapter three.

131. A prominence enhanced by editorial selection (and also by historiographical choices).

132. Indeed, from the region (see Jördens 1999, 126–28).

133. So Wickham 2005, 247 (relying on research by Banaji and Sarris).

134. For "inherent fiscality," cf. Sirks 2008, 131–32.

135. For the "rent roll" (*apaitēsimon*), cf. Rowlandson 1994, 499, and Jördens 1999, 138–39 (somewhat overstated). I am skeptical about Mazza's suggestion (2007, 439) that leases were kept in the countryside (following Rowlandson 1994, 498) or elsewhere in Egypt (e.g., Alexandria). In any case, the fossilized *lēmmata* (receipts) of the accounts of *pronoētai* (estate stewards), which were based on the *apaitēsima*, speak against regular revision of the rolls (pace Mazza 2007, 441). For *relocatio tacita*, the implicit renewal of a lease, cf. *CJ* IV.65.16 (260).

Alongside these spatial lacunae are temporal ones. Though the family and its holdings are attested over the space of 190 years, the bulk of the papyri cluster in the period between 540 and the Persian occupation.[136] Accordingly, it is impossible to follow the rise of the estate under Strategius I and his successors in any detail or, for example, to witness the effects of Anastasius' confiscation (if it in fact occurred).[137] The estate that the extant papyri reveal is the mature *oikos,* with its administrative apparatus well formed.[138] But it is certainly not a monolith; changes may be observed over these eight better-documented decades. It is clear, for example, that the Apiones' (target) rural receipts (including taxes) increased about 30 percent between ca. 542 and ca. 587,[139] and it seems to have been thought necessary to add an extra layer of management, the *antigeouchos* ("landlord's agent"), at some point before the seventh century (probably during the unstable period after Apion II's decease).[140] In conjunction with such developments, we can glimpse shifts in landownership: the incorporation of the property of the indebted Diogenes into the estate[141] and the absorption of some rural settlements and loss of others.[142] The resulting sense of motion over time, alongside the historian's ability to discern (or, rather, generate) patterns, becomes problematic, however, if it ends up suppressing the truly substantial gaps in the data; unlike other assemblages of papyri, the Apion dossier, at least as it currently stands, is actually antithetical to narrativization. This is not a reason to despair or to "wait for all the texts to be published," but there is cause for proceeding with an extra dose of care and self-consciousness.

---

136. Cf. Ruffini 2008, 122, table 10.

137. For ways in which confiscated land might have been handled, cf. Delmaire 1989, 598–610, 620–24. Might *CPR* XIX 6 (probably 515; cf. *APF* 51 [2005]: 288) possibly reference confiscated Apion property in the Herakleopolite?

138. It has been plausibly suggested that imperial methods of estate management were influential on the family; cf. Banaji 1999, 203.

139. See detailed discussion in Hickey 2008.

140. I owe this observation to Nick Gonis. (With the exception of *P.Oxy.* XXVII 2480, discussed below [p. 97], all of the securely dated references to *antigeouchoi* of the Apiones derive from the seventh century.)

141. Vividly recounted in *P.Oxy.* LXIII 4397.

142. Gains: E.g., Piaa (held by Flavius Alexander in 476, by the Apiones in 528/543/558), Mega Eirenarcheion (held by Count John in 503, by the Apiones in 590). Cf. also *P.Oxy.* XIX 2243a.11, 20 (590). Losses: E.g., Aspida (held by the Apiones ca. 565, by Flavius Ioannes in 586). Note also the references to *to dikaion* P(ersonal) N(ame) (discussed below, pp. 59–60), which reveal some information about the evolution of *epoikia.*

Chapter I

Land

The Great Estate

As we have seen, the Flavii Apiones were one of the most powerful families in the later Roman Empire, and it has been asserted that "the basis of the family wealth was their estate . . . in Oxyrhynchus."[1] Indeed, since at least Hardy's *The Large Estates of Byzantine Egypt*,[2] these Oxyrhynchite holdings have been regarded (and served) as the quintessential "great estate"; and in virtual metonymy, they continue to function as a proxy for the larger agrarian "enterprises" of late antique Egypt. One would think that an estimate of the Apion estate's "greatness" would be a prerequisite for this role as archetype, but no one attempted to determine the estate's area until A. H. M. Jones tackled the question in his monumental *The Later Roman Empire*. Jones concluded, "The Apion estates in this district [i.e., the Oxyrhynchite and Kynopolite] . . . would . . . have covered 112,000 arurae, about 75,000 acres or 120 square miles."[3] This amounts to a whopping 35 percent of the total area estimated for these two nomes.[4]

Though he may have been iconoclastic and innovative, Jones's calculations about the size of the Apiones' landholdings certainly did not ruffle any feathers; they were in perfect accord with the then *communis opinio*, that is, the "feudal model" of Bell and Hardy.[5] The large estate is an essential element of

1. *ODB*, s.v. "Apion."
2. Hardy 1931.
3. Jones 1964, 784. He does not attempt to estimate the amount of land devoted to various crops.
4. Bagnall 1993a, 334–335: 323,070 *arourai*.
5. Cf. pp. 5–6 above.

the feudal model.[6] The fiscal demands of the empire, which destroy the decurial class, are its genesis, and imperial weakness allows it to flourish. Populated with its tied proto-serf labor force (the *coloni adscripticii*), it provides the economic support for the centrifugal forces in the provinces; it enables, for example, the usurpation of state functions like law enforcement and tax collection. Since "[t]he feudal model is also by necessity for Egypt an 'Oxyrhynchus model'"[7]—that is, because the Apion dossier is the primary evidentiary basis for the feudal model of late antique Egypt—none of its proponents (then or now) would see any reason to question Jones's figures. Indeed, Sarris has recently reckoned Jones's estimate to be "conservative," because, in his view, it does not include the area of directly exploited land (the *autourgia*), which, he argues, was the profit-oriented portion of the estate.[8]

Jones based his calculation on *P.Oxy.* I 127 (late VI), where it is revealed (ll. 1–2, 7–8) that the Apiones' "glorious house" contributes (*syntelei*) more than 140,618.5 *artabai* for the grain tax (*embolē*) of the Oxyrhynchite and Kynopolite. To arrive at his figure, Jones divided this contribution by 1.25 *artabai* per *aroura*, the rate of the *embolē* attested in *P.Cair.Masp.* I 67057.[9] (The actual result of this equation is 112,495 *ar.*) We can reach a similar conclusion from *P.Oxy.* XVI 1909 (reign of Maurice),[10] which indicates that the total *embolē* for the Oxyrhynchite and Kynopolite was 350,000 *artabai.* The amount that the "glorious house" contributes in *P.Oxy.* I 127 is almost exactly 40 percent of this, and when we apply this percentage to the total area of the nomes (according to Jones, 280,000 *ar.*), we once again reach ca. 112,000 *arourai.*[11] (On the assumption that approximately 70 percent of the area of these two districts was arable, the result is a more "modest" 90,460 *ar.*[12])

These calculations are not problematic in themselves, but Jones's assumptions about the key texts underlying them may be interrogated. Though Hardy had already hypothesized that "the Apion estate was responsible for collecting

6. It is not, however, the causative force in this model that it is, for example, in Weber's "Die sozialen Gründe des Untergangs der antiken Kultur."

7. Keenan 1993, 142.

8. Sarris 2006, 85. His claim about the role of the *autourgia* suggests, of course, that he does not argue for feudalism in the traditional sense; cf. p. 6 n. 21 above. Labor and production on Sarris's great estates were "highly commodifed" (p. 95).

9. The rate in 67057 is, in fact, on the low side; see Zuckerman 2004, 215.

10. Also mentioned by Jones (1964, 1320 n. 28); cf. as well Gascou 1985 (46 and n. 67), which, moreover, provides the rationale for the date of XVI 1909 given here.

11. Using Bagnall's area (see n. 4 above), the result is ca. 129,200 *arourai.*

12. 70 percent: Cf. Rowlandson 1996, 17 with n. 184. Calculation: Multiply 70 percent of the total area of nomes (see n. 4) by the Apion contribution to the *embolē;* divide this product by the total *embolē.*

the taxes of others besides its own tenants,"[13] Jones did not reckon with this possibility. Perhaps he had no reason to question his presumptions, given the extent to which his results were in accord with the scholarly consensus of the time. In any case, criticism came only twenty years later, in the form of Jean Gascou's model of fiscal shares[14]—Hardy's hypothesis expanded and supported. According to the so-called Gascou thesis, *P.Oxy.* 127 represents the glorious house's share of a compulsory public service (or *munus*)—in this case, the collection of the *annona civica* (the food supply for Constantinople).[15] Two-fifths is not an indicator of the Apiones' share of the land in the Oxyrhynchite and Kynopolite; it simply means that for this particular *munus*, the house was responsible for 40 percent of the total tab. No doubt the vast majority, if not all, of the family's land would have been included in this share, but so would the holdings of a great many others who were not among the elites accountable for local administration during the period under consideration. We might compare the "superior Zamindars" of Mughal India, whom Grover indicates "claimed large territorial jurisdictions (*taaluqas*) composed of both their personal *Zamindaries* and those of the petty Zamindars and paid revenues to the state for the entire territory."[16]

The choice between Jones and Gascou is not an insignificant one, a mere quibble about numbers. On the contrary, it strikes at the very heart of the historiography of the late antique Oxyrhynchite (and beyond).[17] If we diminish the great estate, the feudal model is certainly weakened; if we demonstrate that elite participation was willing, some might argue that the model is crippled.[18] To date, the views of neither Jones nor Gascou have managed to become consensus. At present, the revisionist paradigm is in eclipse; in fact, a 2006 commentator, in the wake of Sarris's eloquent critique of Gascou, pronounced the latter's model "largely discredited."[19] My own analysis of data preserved in a still unedited account from the highest level of estate administration (*P.Oxy.*

13. Hardy 1931, 53.

14. Cf. p. 6 above.

15. Banaji (1999, 205) has described the "houses" (*oikoi*) as estates "held in joint ownership and thus immune to the devastating fragmentation of partible inheritance." Of course, the state would also have benefited from their stability, their "corporate personhood."

16. Grover 1964, 274–75 (a reference that I owe to Banaji 2009). A comparative agrarian history of Mughal India and late antique Egypt seems a desideratum, especially in light of Bang 2008.

17. This makes the absence of a discussion of the issue in Mazza 2001 puzzling.

18. "[C]onflict between the imperial authorities and members of the land-owning elite" lies at the core of the arguments in Sarris 2006 (I cite p. 8). But see p. 6 n. 21 above.

19. Mayerson 2006, 100, specifically citing Sarris 2006. Cf. also, e.g., Conor Whately's review of Ruffini 2010 (http://bmcr.brynmawr.edu/2009/2009-07-25.html): "the Gascou line . . . ably challenged recently by Sarris."

XVIII 2196v) suggests, however, that only the Gascou paradigm can stand.[20] Though the text in question is an independent testimonium, distinct from the documentation considered by Jones and Gascou, additional "autonomous" support seems desirable, especially in light of the recent reinvigoration of the great estate model. One way to obtain such corroboration is to analyze estate agricultural production. When we know how much of a crop a certain "sector" produced, we may estimate the area devoted to this commodity and, on the basis of comparative evidence, the area of the estate itself. This is not a simple or quick method but one characterized by the painstaking accretion of carefully sifted evidence from the papyri; we must truly understand a sector before we draw conclusions about (and from) it. A cash crop should be selected so that we might also address recent arguments that the estate was a profit-driven enterprise that made a rational investment in production for the market. Wine is the obvious choice, especially given the emphasis that it has received in the scholarship, but for the sake of thoroughness, the remainder of this chapter will address the market "posture" of every crop attested in the papyri concerning the estate.

Before we proceed, however, some additional words concerning the structures of the estate are necessary to facilitate discussion and analysis. Though the extent of Apion holdings may be a matter of contention, their concentration is not: it is widely accepted that the "mature" Oxyrhynchite estate consisted of properties dispersed over the nome.[21] The family possessed some property in the territory of villages (*kōmai*), but none of these settlements were owned in their entirety.[22] More characteristic of the estate was another kind of community, the smaller *epoikion* (often rendered in English as "hamlet" or "farmstead").[23] Most of these less substantial settlements cannot be localized with precision, so a proper map of the estate is out of the question.[24] At the center of each *epoikion* (figuratively, if not literally) was a building complex that "often, if not normally, took the form of a barrack-block";[25] this housed workers and

20. See Hickey 2008.

21. Cf. Ruffini 2008, 123–24 (and refs.), as well as 243: "Network analysis shows that the Apions acquired property and fiscal responsibility far and wide, without any quantifiably demonstrable regard for the physical proximity of these properties." Prima facie, this seems reasonable for the mid-sixth century, less so for the mid-fifth. Ruffini's analysis does not distinguish between ownership and liturgical ties.

22. Cf. Hickey 2007, 299; Banaji 2009, 84.

23. *CPR* XIX 33, 15n., states, "Ἐποίκιον in the Byzantine period is identical in meaning with κώμη." This was hardly the case.

24. Benaissa 2009 is the latest gazetteer of the Oxyrhynchite.

25. Rathbone 1991, 31 (also see fig. 1, p. 32).

machinery (e.g., processing equipment like presses). Generally speaking, an *epoikion* was owned by a single landholder and had a population in the hundreds (at most).[26] Commentators have compared *epoikia* with the modern Egyptian *ʿizab*.[27]

Alongside references to villages and *epoikia*, we encounter the designations *ktēma* and *chōrion*. The former term is ambiguous: it can reference an enclosed parcel of property—in particular, a vineyard[28]—but in the papyri of the late antique Oxyrhynchite, the meaning "property, holding," often in reference to an *epoikion*, predominates (cf., e.g., "Aurelius Ptolion, the son of Anouthis, his mother being Nonna, originating from the hamlet Ambious, a holding of your extraordinariness)."[29] The meaning of *chōrion*, recently misunderstood by Sarris,[30] is more appropriately discussed in the next chapter.

Epoikia and *kōmai* were grouped into districts called *prostasiai*,[31] each of which was overseen by a *pronoētēs*. A text from 541 suggests a total of sixteen *prostasiai*, but there could have been more.[32] Accounts in various states of preservation are extant for ten of these districts.[33] Five settlements fell within the boundaries of the "average" *prostasia*.[34] *Prostasiai* themselves were grouped into *dioikēseis*, which were overseen by *dioikētai*; there seem to have been as many as five of these larger districts.[35] Falling outside of this administrative structure may have been the *autourgia*, land that the estate exploited directly,

---

26. Single owner: *PSI* VIII 954.26 (from the Apion dossier) would seem to be an exception. Note also pp. 59–60 below. Smallholders within *epoikia:* Cf. *P.Oxy.* LV 3804.92; *P.Warr.* 10. Population: Cf. Ruffini 2008, 109 (table 8), but these calculations are unsatisfying because he does not address the issue of resident (and migrant) wage labor. Cf. also *CPR* X 65.

27. This has been especially emphasized by Banaji (e.g., 1997, 98–99).

28. For this usage, cf. Rathbone 1991, 33. Surely vineyards (though the noun is not modified by the adjective *ampelikos*): *Stud.Pal.* XX 241 (VI–VII).

29. *P.Iand.* III 48.11–14: Αὐρήλιος Πτωλίον υἱὸς Ἀνουθίου μητρὸς Νόννας ὁρμώμενος ἀπὸ ἐποικίου Ἀμβιοῦτος κτήματος τῆς ὑμῶν ὑπερφυείας.

30. In Sarris 2004a, 2004b, 2006.

31. Pace, e.g., Ruffini 2008, 95 n. 3, *pronoēsia* refers to the office (*Amt*); it is not an equivalent.

32. Ruffini 2008, 107. See now chapter 5, n. 50, below.

33. Cf. Mazza 2001, 179. Mazza's *pronoēsia* (*sic*) A is represented by two accounts.

34. Actually 4.8, based on the list in Mazza 2001, 179. Ruffini (2008, 108 n. 67) thinks "that this figure is probably too low," but he misunderstands the terms *dikaion* (on which see below) and, it seems, *ktēma*. Sarris (2006, 30) reports "six or seven." Of course, a district's constituent settlements could change: in 541 (*P.Oxy.* XVI 2032), Maiouma (l. 41) and Pangouleeiou (l. 38) had different *pronoētai*, but in 617 (*P.Oxy.* VI 999), the two settlements were administered together.

35. So Fikhman (1968, 43), though I am not entirely convinced that the same kinds of *dioikēseis* are being referenced in the text that he cites, *P.Oxy.* XIX 2244 (528/543/558). The *dioikēseis* in 2244 do not seem to have a geographical basis (e.g., the *kōmē* Mermerthon appears in the same "district" as the *epoikia* Aspida, Kyamonos, and Leonidou); the matter deserves study. *P.Oxy.* XVI 2031 (VI/VII) certainly has two *dioikētai* of the type relevant for the present discussion.

though at least some of its expenses were charged to *pronoētai.*[36] Wine production and distribution, which were overseen by an *oinocheiristēs* ("wine steward") with estate-wide jurisdiction, largely lay outside the pronoetic bailiwick.[37] The *pronoētai* were responsible only for the collection of gold and grain.[38]

Crops

The extant papyri attest that a variety of crops was produced on estate lands. That these holdings were not exploited in a uniform fashion is clear from even the most cursory examination of the accounts of the *pronoētai.* Cash accounts for a far greater percentage of the (anticipated) inflows in *P.Oxy.* LV 3804, for example, than it does in *P.Oxy.* XVIII 2195: 83.5 percent compared to 46.8 percent.[39] Even within the *prostasiai* themselves, there are divergences. In 3804, the percentages run 100 percent (Apele, Pakiak),[40] 99 percent (Kissonos), 93 percent (Loukiou), 79 percent (Tarousebt), 75 percent (Kotyleeiou), and 65 percent (Trigeou); in 2195, 92 percent (Kainou), 51 percent (Tancheo and Sypheos), 50 percent (Erotos), 48 percent (Tillonos), 47 percent (Amokos), 42 percent (Nekontheos, *dikaion* Kephala), 39 percent (Nekontheos), and 31 percent (Aroures).[41] As a general rule, cereal land paid rents in cereals, while cash rents were assessed on fodder, vineyards (and orchards), and flax.[42] Even if we

36. The income portion of *P.Oxy.* XVI 1913, were it extant, would probably have illuminated the issue of administration. This account, which contains many entries for the (significantly?) autourgic *Exo tes pyles,* also concerned the settlements Iboitos and Kathegetou (see Hickey 1998, 163). The former of these had its own *pronoētēs* (cf. *P.Oxy.* LV 3805.65), so perhaps autourgic land fell under the jurisdiction of the local *pronoētēs.* The organization of *P.Oxy.* XVI 2032 would seem to speak against economic integration on the local (as opposed to estate) level, though *P.Oxy.* XVI 1911.107–38 and LV 3804, 196–212, in which land taken for direct cultivation is credited to the *pronoētēs* (Sarris 2006, 33, has this backward), strongly suggest otherwise. The *autourgia* is revisited in chapter 5.

37. See chapter 4.

38. See Mazza 2001, 134–47, for an overview of other estate personnel.

39. Banaji 2007, 223 (table 2).

40. It should be noted, however, that the account entries are not complete for Apele.

41. For the sake of consistency, these calculations use Banaji's ratio of 12 *artabai* to 1 *solidus* (see Banaji 2007, 223–24 n. e), which reappears in the unpublished verso of the fiscal account *P.Oxy.* I 127 and elsewhere in estate documentation. They exclude sums that are not relevant to the question of the character of the land (e.g., rent paid for a bakery).

42. As Jördens (*P.Heid.* V, p. 227) notes, though, wine was often sharecropped: "In viniculture . . . share contracts were generally the usual contract form." This was not just the case in late antique Egypt; cf. Bloch 1966, 147: "What is certain is that throughout the Middle Ages the institution of *métayage* was either completely unknown in many regions or was normally applied in practice only to particular categories of land, notably vineyards." Concerning flax, cf. Rowlandson 1996, 241 (who also notes the possibility of rents in tow, i.e., coarse fibers of flax).

allow for exceptions, the rural landscape around the *epoikion* Aroures no doubt varied significantly from that around the *epoikion* Pakiak.

What should we make of these variations? Environmental factors might be suggested as an explanation, at least a partial one—the *epoikion* Aroures was in the northern part of the Oxyrhynchite nome, while Pakiak was in the south—but Sarris has recently proposed that we are seeing deliberate specialization of production, one indication (of several) of the commodification and monetization of the "great estates" of Byzantine Egypt. "[A]s emerges with particular clarity in relation to the properties of the Apion family," he summarizes, "the aristocratic estates of the late antique east were emphatically non-autarkic enterprises."[43] Though Sarris's flawed presentation of the pronoetic accounts may lead one to question his argument for specialization from the outset,[44] the broader issue of autarky is an important one and merits our attention. The best way to examine the question seems to be to proceed by commodity (or "sector"), the method I pursue in the remainder of this chapter. At the outset, however, I must clarify a critical point: because there are testimonia that an estate purchased goods and services, it does not necessarily follow that it was a market-oriented enterprise.[45] Of fundamental importance is the purpose of such inputs. Do they simply serve to fulfill the consumption needs of the household (*sensu latissimo*), or are they directed toward or supportive of production of a marketable surplus? Similar scrutiny must be applied to outflows; apparent market activity might just be the interchange

---

43. Sarris 2006, 196–97.

44. The following examples were selected because they are easily addressed (all page references are to Sarris 2006):

1. "The land here [Kotyleeiou] is described as having included *considerable* [my emphasis] portions watered by means of irrigational machinery" (p. 33, citing *P.Oxy.* LV 3804.132, which mentions a lease payment for an irrigation machine and its land or *mēchanē*). The payment in question perhaps corresponds to 7 percent of the land of the *epoikion*. (Pakiak in the same *prostasia*, with no grain rents whatsoever, presents a better argument for a high percentage of irrigated land.)
2. "*edaphos* . . . was the standard term for a subdivision of directly managed land" (p. 33). But we find *edaphē* under lease in, e.g., *P.Oxy.* LXIII 4390 (469, an Apion text), *P.Mich.* XI 611 (Oxy., 412), and *PSI* I 77 (Oxy., 551). Tenure is not a defining characteristic of the designation.
3. "The *epoikion* of Pesta, on the other hand, which is also included in the account [*P.Oxy.* XIX 2243a], contributed both wheat and coin and was a centre for milling, the production of tow, and pig rearing" (p. 35). The tow (valued at less than 1 *solidus*) and pig (1 of them) were *purchased* from the estate (ll. 26, 28).

45. Cf. Love 1991, 87–88. Hardy (1931) does not make this distinction (see index, s.v. "Oikenwirtschaft," with pp. 124–26), nor, e.g., do Bowman (1986, 117) and Cameron (1993, 214) when writing of the estate.

(or redistribution) of resources to meet household needs. Fiscal payments do not qualify, of course, as direct market activity, and any "sales" with a state connection should be carefully interrogated.[46]

Grains (Wheat and Barley)

Wheat (*triticum durum* since the later Ptolemaic period) was the preeminent crop of Graeco-Roman Egypt, and large quantities of it were certainly grown by the Apiones.[47] Yet it has long been thought that the glorious house produced little wheat beyond that which was needed to pay its fiscal (and parafiscal) dues and to fulfill its own consumption (including planting) needs.[48] There is nothing in the current estate documentation that suggests that this view should be revised,[49] and, accordingly, wheat should not be classified as a market-oriented sector.

Barley is not well documented in the Apion dossier, but this silence does not allow the conclusion that the cereal lacked importance for the estate.[50] A significant component of the fodder (*capitum*) allowance for certain soldiers and officials,[51] it would have been required by the *oikos* for such rations, and it appears several times in the dossier in this and other fiscal contexts.[52] That the Apiones did not need to purchase all of their barley—because they received it in rents or from their own directly exploited holdings[53]—is clear, but the evidence will not allow for any kind of quantitative judgment.[54] It seems unlikely, however, that the *oikos* would have produced barley for the

46. This is not to deny the existence of "tied trade," on which cf. Wickham 1988.

47. Pure wheat was apparently not sown; cf. Mayerson 2005.

48. See already Hardy 1931, 100. Cf. also, e.g., Gascou 1985, 36 and notes.

49. *P.Oxy.* XVI 1911.208–9 and 1914.5–7 unequivocally attest the sale of small surpluses of *sitos* (just over 21 and 12 *artabai*, respectively) by the *pronoētai*, but the purchaser was the state.

50. As inferred by Johnson and West (1949, 55 n. 41), who, based on the accounts of the *pronoētai*, write that it "was evidently of small importance." Note that it may be concealed in *sitos*, 10 percent of which, according to Mayerson (2005), was barley. The relatively small amount of barley (76.5 *artabai*) contributed by the *endoxos oikos* in *P.Oxy.* XVI 2020 should not be viewed as an indication of its importance on the estate; cf. Gascou 1985, 48 n. 279.

51. For the *capitum*, see Gascou 1989, 292–95.

52. Cf. *P.Oxy.* XVI 2010 (a fiscal sale) and 2046, as well as 1907.7, 8 (where it seems to be adaerated, i.e., commuted to cash). For barley as a component of an *opsōnion* (salary; for a *comes* or count), see *P.Oxy.* XVI 1913.40.

53. Rents: cf., e.g., *P.Oxy.* XVI 2038.14–15 (*epoikion* Orthoniou), XVIII 2195.73 (from a ferryman for the rent of a landing). Directly exploited holdings: possibly *P.Oxy.* LV 3805.97–99. Again, some barley may be masked by *sitos*.

54. The ratio of *arourai* sown in wheat to those in barley was highly variable on Egyptian estates; cf. Bagnall 1993a, 25 n. 55 and refs.

market, when it pursued a "just enough" strategy for wheat, a cereal of higher value.[55]

Chortos

Like wheat and barley, *chortos* (comprising cultivated plants like clover plus grasses) was a fiscal medium.[56] It was required for the horses of soldiers and officials, as well as for the animals of the *cursus publicus* ("public post"); the estate also needed, of course, to feed its own animals when and where forage was unavailable.[57] *Chortos* is the most frequently attested of the pure fodder crops in the dossier.[58] That the Apiones grew noteworthy quantities of it on directly worked land in Notines Paroriou, to the east of the city of Oxyrhynchus, is certain: *P.Oxy.* LV 3804.241–43 attests 64 *arourai* (.18 km^2) of *chortos* there (protected by field guards, *agrophylakes*),[59] and purchases of *chortos* seed for the holdings in question and property at nearby Path Tampemou (also autourgic) are attested in some number.[60] These purchases were made by specialized agents (the *chortoparalēmptai*),[61] and *pronoētai* throughout the Oxyrhynchite contributed funds for them.[62] The obligation of several workers (*ergatai*), residents of other *epoikia*, to sow *chortos* at Path Tampemou is also documented.[63] The fact that seed had to be bought probably indicates

55. *P.Oxy.* XVI 2032.87 either represents money owed (by an employee) for an advance of barley or money advanced (to an employee) for the purchase of barley. The latter scenario seems rather more likely and, depending on the purpose of the barley, could represent "open" activity.

56. Description: Schnebel 1925, 211–18. Rathbone (1991, 214) renders it as "hay."

57. Natural pasture was rare in the Oxyrhynchite according to Rowlandson 1996, 21.

58. *Chlōra*, "green fodder," hardly appears (cf., e.g., *P.Oxy.* LV 3804.226–27, for the *cursus clabularis*), and green (fresh) hay may be meant by this designation (cf. Rathbone 1991, 233 n. 13).

59. Note also *P.Amh.* II 150 (with *P.Oxy.* XIX 2243a.32), perhaps indicating nonautourgic fodder land in Notines Paroriou.

60. *P.Oxy.* XVI 1911.178–80 (6 *solidi* less 24 *keratia*), 1913.36–39 (total 12 less 48); XVIII 2195.132–33 (5 less 20), LV 3804.244–48 (total 7.5 less 30). (All sums are on the "private standard" [i.e., less 4 *ker./sol.*]; for this and related standards, cf. *P.Oxy.* LXXII 4390, 14–16n.)

61. It is not clear if these individuals worked exclusively for the Apiones or contracted with other landholders as well; I suspect the former. Additional references: *P.Oxy.* XVI 2032.74, XXVII 2480.20–21 (wine for a *chortoparalēmptēs* on a mission); *SB* XVIII 14063 (Philoxenos *chortoparalēmptēs*, homonymous with the individual[s] in *P.Oxy.* 2032 and 3804).

62. E.g., the *epoikia* of *P.Oxy.* XVIII 2195 were in the far north of the nome; those of XVI 1911 (and LV 3804), in the far south. Note also *P.Oxy.* XVI 1913.36–39, which indicates the transfer of part of the liability to another *prostasia*.

63. *P.Wash.Univ.* II 102 (with the correction suggested in Mazza 2001, 131 = *BL* XII 289). The text is not explicit about the crop, but *chortos* is the only product attested at Path Tampemou. The *epoikia* in *P.Wash.Univ.* 102 that can be localized (see Benaissa 2009, s.vv. Πλελω, Μαειουμα) were not located a great distance away.

that the *oikos* was not able to (or chose not to) meet its own consumption and fiscal needs internally;[64] purchases of *chortos,* such as that attested by *P.Oxy.* XVI 1932, would seem to confirm this. The *chortos* in 1932 presumably came from the *epoikion* Loukiou (the *origo* of the seller), a settlement with a very high percentage of cash rents (93 percent of the total land rent expected); accordingly, it might be argued that some of the unspecified cash rents in its pronoetic account represent fodder land.[65] In fact, the part of the nome in which Loukiou's *prostasia* (83.5 percent cash ratio) was situated was a fodder-growing area during the Roman period.[66] Yet the Oxyrhynchite may have been chronically short of *chortos,*[67] since the *chortoparalēmptai* consistently traveled nearly all the way to Hermopolis to get seeds.[68] The *chortos* sector may be classified as "open" to the extent that any purchases supported fixed capital's contribution to the production of marketable surpluses. Regarding the potential of marketing *chortos* itself (given the possibility of scarcity), a "landlord's return on fodder crops was always meagre in comparison with that on cereals."[69]

Legumes, Other Vegetables, and Fruit

Legumes, though a staple of the diet (and an important source of fodder), hardly appear in estate documentation.[70] Since such produce was not a component of any tax, and because these were far less commercial crops than, for example, wine and flax, the silence is hardly surprising.[71] The same may be

---

64. The third-century estate of Appianus was not able to do so; cf. Rathbone 1991, 215. The Apiones' *oikos* received *chortos* through rents (cf. *P.Oxy.* XIX 2243a.54), and it likely was part of the "entry fee" for certain staff positions (cf. *P.Oxy.* LVIII 3952.33–34, *pronoētēs*).

65. Cf. also *P.Warr.* 10.

66. Rowlandson 1996, 18.

67. Rowlandson (1996, 21–22), using earlier evidence, concludes that this was not the case.

68. To the *kōmē* Areos and vicinity, just north of the city; cf. Drew-Bear 1979, s.v. (This assumes that there is not a hitherto unknown Oxyrhynchite village with the same name.) In *P.Oxy.* XXVII 2480.20–21, the *chortoparalēmptēs* travels only to Ptochis.

69. Rowlandson 1996, 238.

70. Noteworthy is *P.Oxy.* LV 3804.251–53, which attests the purchase of 32(?) *artabai* of *arakos* seeds for sowing on some *geouchikai* (here very probably indicating "directly worked") *arourai* and other land (*autourgia* at Notines Paroriou, tenancy on the *ar.* of Apa Apollos?). The *geouchikai arourai* are described as τῶν λιμνασθεισῶν ἐκ δευτέρου (the ed. baldly translates the participle as "watered"; render as "irrigated/flooded for the second time"), and this, combined with the fact that the transaction occurred in the previous indiction but too late to make its account, suggests an extraordinary situation, probably double-cropping. Cf. *P.Flor.* II 194.26–32, as analyzed in Rathbone 1991, 232, 235.

71. Cf. Bagnall 1993a, 27, as well as Rathbone 1991, 381. Cf. also *P.Kellis* IV 96, p. 39.

said of other vegetables and fruits, though these are somewhat better represented in the dossier—primarily because the *endoxos oikos* often leased its lands producing them (typically, it should be noted, for cash rents).[72] This sector's open-market activity could at best be classified as very limited.

Oils

The oil generally encountered in the papyrological documentation is not from the olive, but from the radish (predominantly).[73] A staple of the diet, oil was also a component of the *annona militaris* (military rations), and such distributions account for many of its attestations among the Apion papyri.[74] Leases of estate oil presses, always for cash rents, are documented in the accounts of the *pronoētai* and in the sole extant *apaitēsimon* (rent roll).[75] Two directly operated presses may also be attested;[76] in any case, it is possible that the *oikos* had the right to use the presses that it leased or required lessees to press or even

72. Noteworthy is the *epoikion* Nesou Lachanias πέραν τῆς Ὀξυρυγχιτῶν πόλεως ("opposite the city of the 'Oxyrhyncites"; so *P.Amh.* II 149.5). The Amherst papyrus could indicate that at least part of it was directly exploited, but other documents register rents: *P.Oxy.* XVI 2031.11 (4 *sol.*, private standard) and XIX 2243a.61–65 (34.75 *metron art.* wheat and 6.25 *sol.*, including the *gēdion* of the fruit guards [καρποφυλάκων is clearly preferable to the ed. pr.'s Καρ‹υ›οφυλάκων; cf. now *BL* XI 160] + 30 *sol.* for the *kēpoi* Nesou Lachanias + 4 *sol.* for the *kēpion* Palektoros [cf. XVI 2031.11?]. The lines in question state that these sums have been transferred to another account; they no doubt appear among the *lēmmata* [receipts] at 2243a.1–3.). The *epoikion* Nesou Lachanias also possessed a harbor and a warehouse (cf. *P.Oxy.* LV 3805.35–37).

Additional garden tenancies: *P.Oxy.* XVI 1917.55, 116–17; *autourgia:* probably *P.Oxy.* XVI 1913.67–68 (near the estate house). Orchards: e.g., *P.Oxy.* XVI 1917.99 (lease, 2 *sol.*), XIX 2243a.53–58 (lease, originally 7 *sol.* private standard, now 1.25), XXVII 2478 (surety for a *pōmaritēs* who is an Apion tenant), LV 3805.89–90 (lease, 1 *ar.*, 1 *sol.* + 12.25 *ker.* on the Alexandrian standard). *P.Oxy.* XVI 1913.67–68 refers to orchards (probably exploited directly) around the estate house; cf. also lines 16–18 (and LI 3640), as well as LXX 4789. Date palms: e.g., *P.Oxy.* XVI 2037 + Hickey 1998, 164 (total rent of 7.9 *sol.*), XIX 2243a.27 (rent of date palms and mill = 1.25 *sol.* without *rhopē,* owed by a *koinon* or community), LV 3804.46 (rent of date palms and *synteleia kephalēs* [poll tax] = .5 *sol.*, to be paid by *koinon;* so also line 60, but there 3 *sol.*); *PSI* VIII 954.6 (rent of 750 myriads of *denarii*).

73. Despite being identified simply as *elaion;* cf. Bagnall 1993a, 27–31. For its substance, see Morelli 2004 (against Bagnall 2000). In some texts from the Byzantine Oxyrhynchite it is identified more precisely, as *rhaphanelaion* (e.g., *P.Oxy.* XVI 1837.4) or *elaion rhaphaninon* (e.g., *P.Oxy.* XVI 1870.10). For the pressing of the seed, *lachanospermon*, which furnished the oil, cf., e.g., *P.Oxy.* XVI 1913.65: ὑπὲρ τιμῆς λίθων λαχανοκοπικῶν ἀγορασθέντων καὶ δοθέντων εἰς τὸ γεουκικὸν ἐλαιουργῖον, "for the price of *lachanon*-pounding stones bought and given to the landlord's oil press."

74. The texts are discussed very thoroughly in Morelli 1996, 61–78. Some totals: *P.Oxy.* XVI 1920: 173.25 *sextarii;* 2046: 118.45 *sextarii* extant; XVIII 2196r: 50 *sextarii.*

75. *P.Oxy.* XVI 1917.100 (in kind, but commuted to 3 *sol.*), 2037.30 (the *apaitēsimon;* 4 *sol.*), XVIII 2195.35 (1 *sol.*), 173 (1 *sol.*), LV 3804.77 (2.5 *sol.*); *SB* XXIV 16324.7 (2.5 *sol.*).

76. *P.Oxy.* XVI 1913.65–66, LV 3804.263–66. Note also *SB* I 1975.

purchase its seeds.[77] The *enoikologos*, the employee who managed the Apiones' urban properties, also collected oil for the estate.[78] This oil may have been in addition to a cash rent, but press rents completely in kind are documented elsewhere in the papyrological corpus.[79] The oil received by the *enoikologos* could also have been from estate seeds pressed by urban tenants for the *oikos*. There is no evidence for the purchase of "regular" *elaion* by the estate[80] nor any testimony of direct sales of oil, and despite Pliny's claims,[81] one is left with the impression that the estate's market activity in this sector was limited.[82]

Textiles

Fikhman has asserted that "the textile production of this estate [that of the Apiones] was not even sufficient for its own needs. The administration always found itself forced to buy clothing on the market."[83] Wipszycka, speaking more generally, lends her support to the conclusions of Reil, "who emphasized the importance of production destined to meet the demands of internal consumption."[84] Both of these assessments, which assume that self-sufficiency was a goal, are largely based on silence, and the state of the evidence concerning the Apion estate has changed little since they appeared. A fuller is listed among the "household" staff in *P.Oxy.* XXVII 2480 (l. 278), but this tells little;[85] one might also note the estate's *epoikion* Knapheon.[86]

---

77. Cf. Rathbone 1991, 197–98.

78. *P.Oxy.* LVIII 3958.27–28, called "mysterious" by the editor, presumably in ignorance of VII 1043.

79. In the Byzantine Oxyrhynchite, cf. (probably) *P.Oxy.* LI 3639.

80. *P.Oxy.* LV 3804.229–30 is an acquisition of oil for an artisanal purpose. *P.Oxy.* XVI 2052.4, which may not pertain to the estate, is a purchase of olive oil, *Spanon elaion*, on which see Kramer 1990.

81. *NH* XIX.79: "Whenever they [the Egyptians] are able, they especially desire to sow this [the radish], since it is more profitable than grain and the tax on it is lower, and no plant produces oil more abundantly" (*hoc maxime cupiunt serere, si liceat, quoniam et quaestus plus quam e frumento et minus tributi est nullumque ibi copiosius oleum*).

82. Cf. Benaissa 2006, 7, noting that buyer-creditors in advanced sales of *lachanospermon* almost invariably acquire small amounts, suggestive of purchases "mainly for the consumption needs of the buyer and his family."

83. I am quoting from Wipszycka 1965, 94, which summarizes the Russian-language Fikhman 1962.

84. Wipszycka 1965, 95, referring to Reil 1913.

85. Wipszycka (1965, 138 n. 20) has unnecessary doubts about the nature of this person's relationship to the estate.

86. Cf. Krüger 1990, 275. The reading of τῶν ταπηταρίων (ed. pr: τ̣[α]π̣ι̣δαρ( )) in *P.Oxy.* XIX 2244 fr. 1.1 is a mistake; read τῶν π̣α̣ιδαρ(ίων). (Gonis [2004a, 200–201] has made this correction independently.)

Worthy of consideration, however, is the evidence from other periods. Van Minnen has made a compelling argument for rather significant production of clothing for export in mid-third-century Oxyrhynchus,[87] while medieval al-Bahnasā (the Islamic town), "once the capital of a flax growing district," was "renowned for its textile industry, which was not confined to the manufacture of linen cloth."[88] One could argue that the situation in the late antique Oxyrhynchite was different—for example, that a decline in textile production had occurred with the end of military clothing requisitions,[89] and that the medieval situation was the product of the conversion of grain land to flax production[90]—but there is some evidence that is suggestive of continuity. In the Roman Oxyrhynchite,[91] the villages of Isieion Panga (located in the first *pagus*) and Antipera Pela (Western Toparchy) were centers of flax production;[92] the crop is attested at a few other sites (e.g., the villages of Senokomis in the Western Toparchy and Tampemou in the Eastern),[93] but never farther north than the Themoisepho Toparchy.[94] The evidence for the Byzantine period is rather limited, but it continues to conform to the Roman patterns. Thus in 484 we see a villager from Senokomis paying interest in tow to an Alexandrian *pragmateutēs* ("trader"), an indication, perhaps—given the *origo* of the lender—of the continued importance of Oxyrhynchite production.[95] The same lender reappears about a decade later, this time receiving interest (from a different borrower) in flax (*linokalamē*).[96] Two documents suggest the production of flax on Apion land: in *P.Laur.* III 75 (+ *BL* VIII 165–66; 574), a resident of an

87. See Van Minnen 1986, with the important remarks of Carrié 2004, 30–32 (though it is not clear to me why Carrié believes that the document in question concerns wool and not flax).

88. Goitein 1983, 120. Cf. Udovitch 1999. References in Islamic geographers: e.g., al-Idrīsī (trans. De Goeje and Dozy, p. 58), Ibn Ḥawqal (trans. Kramers and Wiet, p. 157). Note also Rāghib 1991.

89. The majority of the clothing in *P.Oxy.Hels.* 40 (the text analyzed by Van Minnen) is for children, but the papyrus covers only five days. For military requisitions, cf. *P.Col.* IX, pp. 87–105.

90. So Mayerson 1997b, citing the end of the demand from the *annona civica* (grain shipments to Constantinople).

91. Actually, all of the Roman references date to the period 250–375, a span that accords well with the production evidence adduced by Van Minnen (and that possibly indicates a third-century transformation in Oxyrhynchite agriculture).

92. So Rowlandson 1996, 18–19.

93. According to the data in Benaissa 2009.

94. *P.Oxy.* XLV 3254 (312/313/315). For the toparchy of the settlement in this text, cf. *P.Iand.* VII 139.15.

95. *P.Oxy.* VIII 1130, the language of which (cf. ll. 7–8, 20–24) indicates both an existing relationship and the real possibility of its continuance.

96. *P.Oxy.* LXXII 4918, in which he is described as διαπραγματευομένῳ ἐκ πολλοῦ τοῦ χρόνου ἐνταῦθα τῇ αὐτῇ Ὀξυρυγχιτῶν πόλει, "conducting trade for a long time here in the same city of the Oxyrhynchites."

Apion *epoikion* takes a loan of money (with interest in flax) from his *pronoētēs*, while *P.Oxy.* LXXII 4922 (582) documents a similar agreement (with interest in tow) made between another resident of an estate property and a *stippopragmateutēs* ("tow dealer").[97] The settlement in the first of these testimonia was the *epoikion* Netneou, known to be in the former Upper Toparchy, not far from Isieion Panga;[98] the second, the *epoikion* Lithines, was near the village Tampemou.[99] Netneou was also very close to the *prostasia* of *P.Oxy.* LV 3804,[100] which, as noted above, had a remarkably high cash ratio (83.5 percent). Though flax has all but been ignored in previous discussions of the Apion estate, it seems likely that money rents from flax leases contributed substantially to the percentage attested in 3804—an imbalance that is less pronounced, but still significant, at the *prostasia* closest to Lithines (*P.Oxy.* XIX 2243a: ≈ 66 percent).[101] In contrast, the *prostasia* of *P.Oxy.* XVIII 2195, which was situated in the north of the nome where no flax is attested (during any period), has a cash ratio that is nearly 1:1 (46.8 percent); presumably the bulk of these money rents were for vineland.[102] Yet if I am correct that flax was a significant crop on estate land, it is striking that the *oikos*' engagement with this market was *indirect;* we see it shifting risk to its tenants through cash rents.[103] And the

97. Of the five papyri from the Byzantine Oxyrhynchite documenting flax, four are loans with repayment in kind. This is probably not coincidental; such arrangements were likely characteristic of this production sector.

98. Cf. *PSI* III 165. *P.Col.* VIII 238 (e.g., ll. 12–13) attests marshland, perhaps indicative of the water supply most conducive to flax production, near Netneou.

99. The *epoikion* Lithines was near Notines Paroriou, which was part of the same *prostasia* as Tampemou (cf. *P.Oxy.* XIX 2243a.30–31, 35). Senokomis was also in this administrative district.

100. Cf. *P.Oxy.* LV 3804.155.

101. This calculation is based on the adjusted *lēmmata* (ll. 72–73); the figure for the unaltered *lēmmata* (ll. 59–61) is ≈ 65 percent.

102. Cf. appendix B, in which sites connected with estate viticulture are distributed as follows: Northern Zone (comprising the former Lower, Middle, and Thmoisepho Toparchies): 54 percent; Central (comprising the Eastern and Western Toparchies): 27 percent; and Southern (comprising the Upper Toparchy): 19 percent. By way of comparison, Rathbone (1990, 128) estimates that my Northern Zone comprised 60 percent of the area of the Oxyrhynchite; the Central, 25 percent; and the Southern, 15 percent. I myself would put the Northern Zone at 45 percent (by reducing the Thmoisepho to 5 percent and the Middle to 15 percent; for the latter, cf. the criticism of Rathbone in Rowlandson 1996, 18 n. 51) and the Southern at 30 percent (incorporating territory from the old Nome XVII; cf. Bagnall 1993a, 334–35).

103. Flax might also be sharecropped. Indeed, there is an apparent shift from cash rents to sharecropping in the fourth-century Oxyrhynchite, but the sample consists of texts entirely from the dossier of Dioskoros and Leonides, who may have simply preferred share tenancy for flax land; cf. Rowlandson 1996, 247 n. 129.

The estate's role in textile production, if any, is unknown, but if it were substantial, the lack of direct exploitation would be curious.

estate's most significant transaction in flax could have been a purchase (and was certainly not a sale).[104]

The leasing of sheep for cash, perhaps on terms similar to those attested for the (Roman) estate of Antonius Philoxenus, is mentioned twice in the dossier.[105] These are the sole secure references to pastoral activity on the estate.

Viticulture

With wine, we come to the agricultural product most referenced in the estate dossier after wheat. In the wake of this extensive documentation, and given the Apiones' great wealth and viticulture's traditional position at the center of Roman investment ideology and practice,[106] it is not surprising that commentators have long concluded that the family's expenditures on vineyards were substantial. Thus Hardy wrote, "On the Apion estate wine was produced on a large scale,"[107] while Johnson and West noted that Apion II's "vineyards must have been fairly extensive."[108] More recently, Banaji has written of "the stimulus of an expanding wine industry" that "encouraged [Egyptian] producers to structure . . . investments [in perennial irrigation] carefully," and he has supported his statement with examples of such prudence from the Apion dossier.[109]

Though reasonable, all of these assertions are, at the end of the day, impressionistic—the products of glimpses of certain facets, even a single facet, of the extant evidence concerning the estate. Hardy appears to have been influ-

---

104. *P.Oxy.* XVI 2033.19, an *expense* of difficult interpretation. It reads ὁ(μοῦ) νο(μίσματα) ⳨μ [ed. pr.: ϱμ] κ(εϱάτια) ιϛ, κιντ( ) [ed. pr.: Κίντου?] ἐν στιππ(ίῳ) νο(μ.) οβ, (γίνεται) νο(μίσματα) ͵αιβ κ(εϱάτια) ιϛ. κιντ(ηνάϱιον) [*sic*] ἓν στιππ(ίου) cannot be read; the resulting price is absurdly high. κιντ(ηνάϱια) ἐν στιππ(ίῳ) is awkward. A personal name in the dative (an error for Κουίντ(ῳ)?) would indicate a very large payment in kind. I lean slightly toward the second alternative.

The *pronoētēs* in *P.Laur.* III 75 (discussed above) could have been acting on behalf of the estate but is more likely to have been looking after his own interests. The only evidence for an estate sale is *P.Oxy.* XIX 2243a.26, where there is an entry (ὑπὲϱ) τ̣ι̣[μ(ῆς)] στιππ(ίου) (the *tau* does not need the dot) due from the farmers of the *epoikion* Pesta; the sum is a small one, however, less than one *solidus.*

105. *P.Oxy.* LV 3804.93 (identical entry in *SB* XXIV 16324.23): 6.167 *solidi,* inclusive of *synteleia kephalēs* (poll tax). For Antonius Philoxenus, cf. Rathbone 1991, 202–11.

106. Already in Cato, *Agr.* 1.7 (e.g.). This is not to imply an uncontested discourse; cf., e.g., Tchernia 1986, 209–21.

107. Hardy 1931, 118.

108. Johnson and West 1949, 55.

109. Banaji 1999, 206 and n. 80; cf. also, e.g., 2007, 94. For another example, see Syrkou 1995, 132: "On Apion lands, from which a large quantity of documents regarding wine originate, we know that this was produced on a large scale."

enced by the number of locations attested with vineyards or producing wine, for his statement is followed by a list of twenty-four such places. Of course, a large number of vineyards does not necessarily correlate to an immense surface area under vines.[110] Johnson and West refer to a large account roll (*PSI* VIII 953), in which (l. 5) a shade over 36,700 *dipla* of wine is revealed to have been distributed in a single first indiction. Yet this is not firm evidence for estate production but only testimony for expenditure, and the incomplete account attests to the *purchase* of 2,000 *dipla* elsewhere (ll. 72–74).[111] In Banaji's case, we need not attribute the extension of irrigation technology, the timing of which is itself an uncertain proposition,[112] to a stimulus from the market for wine. Though the spread of perennial irrigation would certainly have enabled landowners like the Apiones to increase their grape production, it seems unlikely that we can decouple technological developments from pressures resulting from Diocletian's fiscal reorganization.[113]

Glimpses and impressions may seem part and parcel of the ancient historian's trade, but the evidence for viticulture on the Apion estate is such that something different is possible—indeed, has long been possible. The extant papyri furnish more than enough data for the quantification and evaluation of the estate's investments in viticulture.[114] This is historiographically fortuitous, for viticulture is also the field on which must be fought the struggle over the economic character of the Apion estate—and, by virtue of metonymy, that of the "great estate" of late antiquity; as the preceding survey of agricultural production has demonstrated, other sectors with significant market potential were largely characterized by autarky and risk avoidance, and it is unlikely that unpublished papyri concerning the Oxyrhynchite estate will alter this picture.[115]

---

110. On the contrary; cf. Ricci 1924, 15–16: "These are mostly small vineyards, the areas of which vary from 2 (sometimes 1 to 1 1/2) to 5 *arourai* as a result of the large fragmentation of farmland across Egypt, and of land cultivated with vines in particular." Ruffing (1999, 234–63) has more data at his disposal and presents a more nuanced picture, but the conclusion that the vast majority of Egyptian vineyards were small remains intact.

111. Note also line 13 of this text.

112. Cf. Hickey 2008, 100.

113. See, e.g., Oleson 1984, 379–80; Rowlandson 1996, 63–65; and even Banaji 1992, 227.

114. The term *quantification* is used loosely here; the demarcation of quantitative limits is a more accurate description.

115. This is an informed hypothesis. I do not exclude the possibility that the estate's nonagricultural activities were significant. The urban sector is poorly attested, though there is documentation that the estate farmed out the collection of its city rents for the considerable sum of 125 *solidi* (*P.Oxy.* LVIII 3958, 614); Ruffini (2008, 117) notes that this could indicate hundreds of urban properties. There is deafening silence concerning agricultural units beyond the Oxyrhynchite and the impact of activities like the luxury trade (for the staggering sums at stake in the Roman period, see Rathbone 2001).

If the Apion estate was indeed "a curious prefiguration of something intrinsically modern" and "highly commodified" with respect to its labor force and production, its viticultural sector will demonstrate this.[116] "Der liebe Gott steckt im Detail," however, and we must begin by venturing down into the morass of the papyri to construct the meanings of fundamental terms and concepts.

116. Quotes: Banaji 1999, 213; Sarris 2006, 95. Sarris seeks a way off the inevitable battlefield through his "bipartite" estate (defined at 2004b, 64, e.g.), which owes its inspiration to Kula 1976 (among others). I revisit this issue in chapter 5.

Chapter 2

Vineland

To describe and quantify viticulture on the Apion estate, we must first, of course, identify it in the sources. At the outset, I should note that no complete listing of Apion vineyards from any period of the estate's history has been published; this information must be painstakingly culled from the dossier. Our principal testimonia for viticulture in the *oikos* are various kinds of agreements (in particular, irrigation machinery receipts and advanced sale contracts), receipts documenting wine deliveries to the estate and distributions by it, and accounts. In very many of these texts, terminology presents a difficulty; indeed, the precise meanings of some important designations have eluded previous commentators. Thus one goal of the present chapter is to establish a firmly grounded vocabulary for discussion of the issues surrounding estate viticulture. But terminology is only the beginning of our impediments. If, for example, we accept the "Gascou thesis," the status of every property mentioned in an estate text must be scrutinized to determine which of the following ownership scenarios pertain: Apion property, imperial property administered by the Apiones,[1] or property that fell within the Apiones' share of a fiscal liturgy (i.e., property for which the estate collected taxes). Even if one chooses to reject Gascou altogether, the possibility that not every property mentioned in an estate text belongs to the *oikos* still requires recognition. This is especially true in village contexts, since no landowner owned these larger settlements in their entirety.[2] The second objective of this chapter, then, is to provide some clarity

1. As noted in the introduction (p. 9), the family's involvement with the management of imperial property may be traced all the way back to the beginning of the documentation.

2. Indeed, villages probably functioned as synapses between the estate and the outside world.

(if not solutions) to issues of ownership. A list of sites associated with estate viticulture (informed by the findings herein) appears in appendix B.[3]

Terminology

General Terminology in the Byzantine Oxyrhynchite

Ampelos is the word commonly employed to denote vineland,[4] though it continues to appear with the more classical sense of "vine."[5] The adjective *neophytos* is used to describe newly planted vineyards;[6] both *phyteuō* and *kataphyteuō* are employed for the act of planting.[7] Neither *emphytos* ("planted [with no impediment to use]") nor *symphytos* ("fully cultivated [not neglected]") has appeared in Byzantine papyri from Oxyrhynchus (in description of vineyards or anything else). *Chersampelos,* or dry vineland, which was often converted to grain use, last appears in the Oxyrhynchite documentation in the earlier part of the fourth century.[8] One does, however, encounter concessions in administrative accounts of the Apiones for vineland that is described as ἔνχορτος (read ἔγχορτος, "under grass").[9] Since vineyards required perennial

3. Needless to say, the fragmentary nature of our evidence renders it incomplete.

4. Cf., e.g., *P.Mil.* II 64.6; *P.Oxy.* XVI 1899.11, 1900.13, 1915.4, 1982.11, 1990.18 (see *P.Oxy.* LXX, p. 141), XXXVI 2779.11, LV 3803.8, 3804.223; *PSI* VIII 881.5; *P.Select* 20.7; *SB* XXIV 16312.17.

5. Cf., e.g., *P.Oxy.* XVI 1910.13 (with Hickey 1999, 325–26 = *BL* XII 142), LV 3804.176.

6. Cf., e.g., *P.Oxy.* XVI 1911.175 and 1912.152, both of which refer to reeds acquired for *neophyta chōria* (*chōrion* is discussed in the next section of this chapter). I suspect that the "young plants" that are surrounded by a wall in *PSI* VII 809.5 are vines (cf. Schnebel 1925, 242–44), but *neophytos* can also be used for other plants, e.g., olive trees (cf. *P.Mich.* IX 554.9 and *P.Ryl.* II 138.9). The *neophyta chōria* in *P.Oxy.* LXVIII 4704.3 (626) probably also included fruit trees (pace the ed.), since this text concerns the land around the former estate house (*proastion*) of the Apiones, cf. Mazza 2001, 85–86. For the botanical contents of a vineyard, cf. *P.Paramone* 18, 21n. For an *epoikion* Neophytou, see *P.Oxy.* LVII 3914.6. Note also *P.Oxy.* LXVIII 4702.7 (*epoikion* Neophytou Antiochou) and *P.Select* 20.3 (*epoikion* Neophytou Banou); an *epoikion* Neophytou Hierakos appears in the forthcoming dossier of Philoxenos *oinocheiristēs*. *SB* I 1973.20 probably read "*chōriou* N/neophytou"; given the meaning of *chōrion*, the common adjective is tempting, but this sherd may well be from a wine jar, with its text indicating the vineyard of origin (obviously not just planted in that case).

7. *phyteuō: P.Oxy.* XVI 1910.13 (with Hickey 1999, 325 = *BL* XII 142), 1911.104, XIX 2243a.48, LV 3804.176; it is used in reference to *akanthai Aigyptiai* in XIX 2239.14. *kataphyteuō: PSI* VIII 881.8 (presumably in reference to vines).

8. See *P.NYU* II 40v.18. *P.Oxy.* XVI 2038.20 (VI/VII) is a concession to a community (*koinon*) of vineyard specialists *hyper abrochou,* but the payment conceded was to have been made in grain (as one might expect given the presence of *abrochos*, i.e., land that was *uninundated* and therefore was probably not vineland).

9. E.g., *P.Oxy.* XVI 1911.91, 1915.7, LV 3804.163.

irrigation, it is not surprising that they are often associated with irrigation machinery (*mēchanai, zygon*).[10]

τὸ χωρίον

In the late 1990s, Roger Bagnall and I, working independently, reexamined the term *chōrion.*[11] Bagnall, who studied the references in Preisigke's *Wörterbuch* that had been cited as meaning "Ackerstück, Landstück, Grundstück, Landgut, Landbesitz," noted, "[F]rom (at the latest) the first century AD the reference to uninundated land used for tree crops is clear" (1999, 330–31). He further observed that the supposed transformation of the meaning of the term in late antiquity, to a synonym for *kōmē* (village),[12] had not been demonstrated, concluding that "the dominant sense was that of non-inundated land, primarily used for growing tree crops of various sorts, but especially vineyards, and sometimes also containing farm buildings or equipment" (332).[13]

My own investigation was limited to the usage of the term in the Byzantine Oxyrhynchite. Like Bagnall, I found no support—beyond non-Egyptian sources like the Novels of Justinian or the later *Nomos geōrgikos* (*Farmers' Law*)[14]—for the notion of a late antique transformation in meaning. To be sure, a change was apparent in the papyri dating from after the Islamic conquest,[15] but it seemed unsound to extrapolate such back to the period before

10. *mēchanai,* e.g.: *P.Mil.* II 64; *P.Oxy.* XVI 1899, 1900, 1982, XIX 2243a.48, XXXVI 2779, LXX 4781, 4785, 4788; *P.Select* 20. *Zygon: P.Oxy.* LV 3803. For the use of *natural* irrigation in *Fayum* vineyards, see Rathbone 1991, 222.

11. Bagnall 1999; Hickey 2001, 30–34.

12. With regard to this issue, Bagnall cites Drew-Bear 1979, 42 (which, tellingly, references Louis Robert and Paul Lemerle, not any papyri). Cf. also, e.g., Krüger 1990, 44.

13. Here it may be noted that the *chōria* in *SB* XXVI 16351.6 (IV) should probably be understood as "vineyards," not "Dörfer [villages]"; the same may be said of *SB* XX 14402.1.b (V/VI) and 14403.2 (512/527/542). In addition, "vineyard" should be substituted for "estate" in the translation of *BGU* XVI 2632.7 (after 7 BC).

14. For the *Farmers' Law,* see the overview in Wickham 2005, 462–64.

15. A telling example is *P.Lond.* IV 1354.9 (Aphrodito, 710), where one reads, καὶ μὴ ἀπαιτήσῃς τοὺς τῶν χωρίων σῖτον καὶ ἄρτον, "And do not exact the grain and bread of the *chōria* [i.e., villages or, rather, fiscal/administrative units: cf. *CPR* XXII 1, 6n.]." See also *CPR* VIII 74 (698), an *entagion* (demand note) addressed τοῖς ἀπὸ χωρίου Δικαίου τῆς Ἀρσινοίτου παγαρχίας that only concerns grain. For comparable usage in the Oxyrhynchite, see, e.g., *P.Leid.Inst.* 80 (VII–VIII), *Stud.Pal.* X 198 (VII–VIII; HGv's "Arsinoites?" is incorrect); cf. also *P.Mich.* XV 748.4 (651/666/681; cf. Gonis 2004a, 198–99). Note, however, the opposition in *P.KRU* 65.58 (2nd half VII).

Sarris (2004a, 289), citing Banaji (2007 [2nd ed.], 175), states that *chōrion* is a synonym of *epoikion* ("hamlet, farmstead," a term used for the Apiones' estate settlements) during the Byzantine period, but Banaji is speaking of the Islamic period. This leads Sarris to make statements like the following this major transition.[16] None of the papyri of the Byzantine Oxyrhynchite were found to have a usage of *chōrion* that did not correspond to Bagnall's "dominant sense." In fact, while I would not exclude the possibility that *chōrion* had other meanings,[17] "vineyard" clearly should be the *point de départ* when it is encountered in late Oxyrhynchite papyri.[18] I here provide some examples in which the proper rendering is clear. In *P.Oxy.* XVI 1833.4 (late V), a certain Ioannes tells a notary to find out τὴν ποσότητα τοῦ οἴνου τοῦ παντὸς χωρίου;[19] the precise meaning of *chōrion* is no doubt confirmed three lines later when we encounter γῆν ἢ καὶ χωρίον—a combination that is immediately reminiscent of the earlier χωρίον . . . καὶ σιτικὴν γῆν (*P.Oxy.* XXXIV 2712.6–7; 292/93) and ἀμπελικὸν χωρίον καὶ σειτικὰς [read σιτικὰς] ἀρούρας (*P.Oxy.* VI 907.8; 276), not to mention the contemporary εἰς ἄμπελον καὶ εἰς ἀρόσιμον γῆν, "to vineyard and arable land."[20] Similarly, it seems hard to avoid rendering *chōrion* as "vineyard"—in fact, Grenfell, Hunt, and Bell do so—when it is paired with orchards, *pōmaria,* as an object of irrigation in *P.Oxy.* XVI 1913.3–4 (555?).[21] The *neophyta chōria* for which new reeds are purchased in both *P.Oxy.* XVI 1911.175 (557) and 1912.152 (ca. 555) are clearly vineyards.[22] In *P.Oxy.* XLVIII 3430 (dated to IV) a *chōrion* is de-

(2004a, 302): "The ubiquity of this form of estate structure [his "bipartite estate," for which see Sarris 2004b, 64] in the late antique east is perhaps suggested by the fact that one of the Greek terms used in relation to the settlements of the great estates—*chorion*—was to come to serve as the standard medieval Greek word for 'village.'" Mazza (2001, 80), in contrast, knows (and follows) Bagnall 1999. Bonneau (1983, 514 n. 28) recognizes the Islamic transformation in meaning but provides a curious definition for the earlier period: "a subdivision of part of the alluvial plain (*pedion*) able to be irrigated by submersion."

16. Cf. Banaji 2007, 156–57.

17. It is rendered in Coptic by at least three different words: ⲉⲓⲱϩⲉ, "field"; ϯⲙⲉ, "village"; and ϭⲱⲙ, "garden, vineyard, property." It would be interesting to determine if these usages have a chronological component.

18. Schnebel (1925, 242) seems to indicate that the adjective *ampelikon* should be present for *chōrion* to mean "vineyard," but *chōria ampelika* do not appear in the Oxyrhynchite documentation after the fourth century. They are attested in other Byzantine papyri, however, e.g., *SB* XXII 15606.5, 7 (VI, Hermopolite).

19. Eds.: "the amount of wine from the whole *vineyard* [my emphasis]."

20. This phrase and its variants appear in receipts for parts of irrigation machinery. *P.Oxy.* XVI 1982.11 (497) has been cited here.

21. *P.Oxy.* XVI 1988 (full edition in *P.Oxy.* LXX, pp. 137–38) concerns an irrigation machine called "Of the *Choria*" that supplies water to "the same *chōria.*" The editors have cautiously rendered *chōrion* as "plantation"—an imperfect choice given its connotations of the agricultural regime (and chattel slavery) in the American South. It is virtually certain that tree crops (requiring perennial irrigation) are at issue.

22. Cf. the earlier (296–307 or 324–30 < *BL* VIII 230) *P.Oslo* III 146.6–8, a payment of wine to some vineyard specialists (*ampelourgoi*) πρὸς ἀνανέωσιν τῶν νεοφύτων, "for the restoration of the young plants." For reeds in viticulture on the Apion estate, see also pp. 72–73 below.

scribed as derelict; it has gone εἰς ἔρημον (ll. 5–6). The editors translate *chōrion* with "plot," but the meaning is certain from the reference to the ἀμπελουργίαν τοῦ χωρίου, the vine dressing (or vineyard work) for the *chōrion,* in line 9. Finally, in a document from the mid-fourth century (*P.Wisc.* I 12; 345), a *hydroparochos* (irrigator) vouches that an *ampelourgos* (vineyard specialist) will attend τῇ καλλιεργίᾳ τοῦ αὐτοῦ χωρίου (ll. 9–10), "to the good cultivation of the same *chōrion.*"

There are numerous occurrences that are somewhat less certain but for which the meaning "vineyard" seems probable. *Chōrion* is associated with wine in *P.Col.* VIII 239 (dated to IV), an account of the ῥύσις (literally, "flow") χωρίου Φνᾶ.[23] It appears in similar contexts in *P.Oxy.* XIV 1735 (also dated to IV) and XVI 2043 (dated to V). The *ampelourgoi* of the χωρίον προαστίου are provided vessels for the *katangismos* of wine in *P.Oxy.* XVI 1913.34–35,[24] and one encounters an *ampelourgos* from the χωρίον Λευκίου in *P.Mich.* XIII 674 (VI).[25] In an advanced sale of wine from the seventh century (*PSI* X 1122), the seller-debtor agrees to take the empty wine jars (*ta koupha*) that the buyer-creditor has provided ἕως τοῦ ἐμοῦ χωρίου, "to my *chōrion*" (ll. 26–27).[26]

Remains of late amphorae from the site of Oxyrhynchus very likely support the interpretation of *chōrion* presented above. Grenfell and Hunt report that the "Byzantine amphorae often had two names preceded respectively by χῶρο(ς) [*sic*] or μέρος,"[27] and note "a quantity of Byzantine pieces with the formula χωρ(ίου), followed by a place-name, and μέρ(ους) or μερ(ίδος), followed by a personal name."[28] *Chōrion* here probably indicates the vineyard of origin, the wine's "appellation."

On examining the Byzantine documentation, one might cite in objection *P.Oxy.* XVI 1910.12 (VI/VII): συνεχωρήθησαν τοῖς ἀπὸ Λεωνίδου ὑπὲρ τῶν

23. For *rhysis,* see Kruit 1992b. Cf. also the earlier *P.Oxy.* XVII 2153 (III), which gives the ῥύσις χωρίου Διονύσου.

24. *chōrion proastiou:* For this place, see appendix B, s.v. ΠΡΟΑΣΤΙΟΥ, and Mazza 2001, 84–87. *katangismos:* "Bottling" is obviously not appropriate; translate as "putting in empty vessels [*koupha*]" or the like.

25. For this place, see appendix B, s.v. ΛΕΥΚΙΟΥ. If the papyrus has been dated too early, *chōrion* may mean "fiscal district" here.

26. Here again *chōrion* may have its later (postconquest) meaning; cf. Gonis 2005, 171. For sales in advance of delivery and their logistics, see Jakab 1999 and (more fully) 2001. Possibly relevant to this discussion of *chōrion* is *O.Ashm.Shelt.* 39.4 (VI, perhaps of Oxyrhynchite provenance): *staphylēs* ["grapes"] *chōriou.*

27. Grenfell and Hunt 1902/3, 7. χῶρο(ς) is an error for χωρ(ίου) or the like; cf. the references in the next note. For *meros* (and *meris*), see the next section of this chapter.

28. Grenfell and Hunt 1904/5, 15. See also Grenfell and Hunt 1905/6, 15, and Grenfell and Hunt 1906/7, 11.

χωρίων κατὰ τὸ ἔθος σίτου καγκέλλῳ ἀρτ. μ, "conceded to those from Leonidou for the *chōria,* according to custom, 40 *cancellus artabai* of grain." If the *chōria* in question were vineyards, why would the concession be in wheat? Unfortunately the *lēmmata* ("inflows" or "receipts") for this account are summary, and in any case, the account concerns wheat alone.[29] But it is entirely possible that these 40 *artabai* were conceded because the lands in question no longer produced grain. κατὰ τὸ ἔθος is certainly consistent with a major change like that from arable land to vineland, and the *epoikion* Leonidou is attested as a wine-producing area.[30] An explicit concession for such a conversion appears in the following line (which concerns another property).

The term ἀπότακτον χωρίων, which appears with some frequency in the accounts of the Apiones, is discussed in detail below.

ἡ μερίς

The abbreviation μερ( ) appears in wine accounts and other texts from the Byzantine Oxyrhynchite, and the word intended (μερίς or μέρος) is not always readily apparent. Grenfell, Hunt, and Bell believed that *P.Oxy.* XVI 2043 (V), in which μεριδ( ) appeared, resolved the matter,[31] but a solution is not that simple. Wine was commonly sharecropped, and despite the fact that not a single sharecropping lease from the Byzantine Oxyrhynchite has been published,[32] there are several accounts in which *meros*—the term typically employed to denote the quota[33]—might well be read. *P.Oxy.* XVI 2051r is one such document. The first column of this text contains entries in *arourai,* each of which is followed by γεουχ( ) μερ( ) and a quantity of *dipla.*[34] γεουχ( ) no doubt stands for a form of γεουχικός, and Grenfell, Hunt, and Bell, presumably based on *P.Oxy.* XVI 2043, have resolved μερ( ) as μερ(ίδος), yielding γεουχ(ικῆς) μερ(ίδος). Yet the obvious interpretation is that this tag means "landlord's share," that is, the sum received by the *geouchos* from the immedi-

29. Vineland is, however, mentioned; cf., e.g., line 13 (with Hickey 1999, 325–26 = *BL* XII 142).

30. For the *epoikion* Leonidou, see appendix B, s.v. ΛΕΩΝΙΔΟΥ *P.Oxy.* LXX 4797.6–7, in which an irrigation machine named "Of the Southern *Chorion*" only supplies water to arable land, may reflect the opposite situation, i.e., a vineyard that had been converted to grain production.

31. *P.Oxy.* XVI, p. 264: "The papyrus shows that μερ( ) where used of vineyards is to be expanded μερ(ίδος) rather than μέρ(ους)."

32. Jördens (*P.Heid.* V, pp. 228–29) suggests that the sharecropping agreement had been replaced by the "lease of work" (μίσθωσις τῶν ἔργων). For further discussion, see chapter 3.

33. E.g., *P.Giss.* 56.17–19.

34. Pace the editors, the document has only two columns: columns i–iii and iv–xi of the editio princeps. Though both columns concern viticulture, they do not seem to be related, and an example of scribal practice seems likely, all the more so given that the papyrus is a palimpsest.

ately preceding *arourai*. Though *meris* conceivably could have the appropriate sense, *meros* would certainly be expected.[35]

γεουχικ( ) μερ( ) also appears on the ostrakon *SB* I 1969 (VI),[36] and in this receipt there can be little doubt that a form of *meros*, "share," was intended.[37] Incidentally, this text also has the abbreviation μεριδ( )—which suggests that we might assume that μερ( ) resolves to a form of μέρος unless there are compelling reasons to think otherwise.

When *meris* does appear, its sense is not always clear. Drew-Bear, in her gazetteer of the Hermopolite nome, has noted,

> It depends on the context and also the period. Under Roman rule, *merides* can represent the subdivisions of the territory of a village and in that case they typically are numbered . . . They also can designate the divisions of a large estate . . . Sometimes they bear names of localities and seem to correspond to fiscal divisions within a toparchy; they often are connected with the presence of granaries . . . In the Byzantine period *merides* are generally distinguishable by personal names. According to H. Maehler, these belong to financial officials; their names serve to distinguish the divisions for which they are responsible, and these designations can continue in use [after the departure of the official in question].[38]

Drew-Bear then mentions Gascou's fiscal interpretation, which Palme has summarized well.

> It is not a matter of territorial subdivisions of fiscal districts, but rather abstract concepts . . . These are "shares" [*merides*] of the total tax revenue, which originally (probably in the 4th century) had fallen to a large landholder and then continued to exist as collection units under their old names, in some cases into the 7th century.[39]

Drew-Bear seems to imply that the fiscal sense of *meris* superseded other meanings in the Byzantine Hermopolite. Was this also the case in the Byzantine Oxyrhynchite, specifically in those cases where it appears in the context

35. Cf. also Text I (abstracted in appendix A). Reference to a cultivator's (half) *meros* is made, e.g., in *P.Kellis* IV 96.1772.

36. For the reading, see *BL* II.1 19.

37. Cf. *SB:* μέρ(ους?).

38. Drew-Bear 1979, 43–44.

39. Palme 1994, 53. Palme is speaking of Hermopolite practice, but the description is equally applicable to the contemporary system of *oikoi* and *merides* at Oxyrhynchus. See Gascou 1985, 40–49.

of wine and/or viticulture? Packman, the editor of *P.Wash.Univ.* II 105 (VI–VII), appears to think so; regarding a prefatory message concerning a shipment of wine and related products from the *meris* Mataei to the *geouchos,* she writes, "The items forwarded will have been part of a system of exactions collected by the landowner for the state, with μερίς being the proportion assigned to a particular place."[40] This argument is not compelling. Though wine certainly was levied and deployed for state purposes,[41] the milieu of this particular text seems to be rather more private than public; beyond the presence of *meris,* there is little to suggest a fiscal context.[42] In truth, Packman's statement reveals a fundamental misunderstanding of fiscal *merides,* which were assigned not to places but to people—or, more precisely, to institutions like the *oikoi.* For her interpretation to stand, Packman would have had to have argued that Mataei is a person and that ἐκ τῆς μερίδος Ματαεί has a partitive sense. Packman does not argue such, however, and in fact notes that Mataei is more commonly associated with places (*epoikia, chōria,* and *mēchanai*).[43] It seems unlikely that ἐκ τῆς μερίδος Ματαεί has anything but a local sense.[44]

An examination of the relevant texts suggests that Oxyrhynchite viticultural *merides* simply represent divisions—perhaps ones demarcated concretely[45]—of a vineyard or *chōrion.* The context is clearly private in *SB* I 1969 (VI), which records the *geouchikon meros* ("landlord's share") from the *meris* of the heirs of Andreas,[46] a farmer (*geōrgos*) from the *epoikion* Phatement;[47] these heirs were probably responsible for the exploitation of part of the vineyard. *P.Oxy.* XVI 2042 (V) records quantities of wine from various *ktēmata* that have been carried on the *libernos* (liburnian) of Phoibammon the *nautēs* (boatman). The jars provided by οἱ ἀπὸ Δάφνου are said to be "for the 5 *merides.*" The context—*merides* are not mentioned elsewhere in the text—suggests that "the 5 *merides*" is an actual place, probably a vineyard somehow divided into five sections (at least at one time). In another wine account, *P.Oxy.* XVI 2043 (V), one finds entries for the *meris* Lamasonos, the *meris* Paulou,

---

40. *P.Wash.Univ.* II, p. 201. The editor renders *meris* as "share." κρ( ) in line 1 of this text should perhaps be resolved κρ(ατήρια); cf. *P.Oxy.* XVI 2049 (VI).

41. See chapter 4.

42. The use of a *symmachos* for the delivery is ambiguous; cf. Jördens 1986a, 105–11.

43. *P.Wash.Univ.* II 105, p. 201.

44. Similar misinterpretations appear in *P.Col.* VIII 239 (cf. 2n.) and X 284 (cf. 6n.).

45. I note the tendency to wall vineyards; see (generally) Schnebel 1925, 242–44. Among the papyri linked with the Byzantine Oxyrhynchite, only *PSI* VII 809.5 (VI) is relevant, and its provenance is not certain.

46. As noted above, the abbreviation is an unambiguous μεριδ( ).

47. For this place, see appendix B, s.v. ΦΑΤΕΜΗ(Ν)Τ.

and the *meris* Ariou beneath the heading for the *chōrion* Petroniou.[48] The two other *chōria* in this account have no such subheadings, and, in any case, it seems most improbable that the responsibility for the fiscal charges on a single *chōrion* would have been divided between three parties. Much less strained is the explanation that the *chōrion* Petroniou was divided (at least at one time) between a Lamason, a Paulos, and an Arios.[49] A fiscal context is a similar stretch for *P.Oxy.* XIV 1735 (IV), which includes the yield (*rhysis*) from the *chōrion* Deios. There are subheadings for the *merides* of Ioannes and Serenos; must from the latter was produced at a second *lēnos*.[50] This text also suggests that there may have been some onomastic fossilization: the jars from the *meris* of Ioannes are delivered not by a Ioannes but by the *ampelourgoi* Aion and Choous.[51]

Three papyri should remove any lingering doubts. In the first, *P.Oxy.* VI 940 (V), an unknown administrator writes to Ioseph the *notarius.* He opens his letter with this statement: συνορῶ τέως ἐν ταυτότητι μεῖναι τὰς ῥύσεις ἄχρι τῆς τελευταίας μερίδος ἵνα μὴ δόξωμεν διώκειν τοὺς ἄλλους τοὺς μήπω τρυγήσαντας.[52] The context is almost certainly private.[53] The *rhyseis* from these *merides* are not distinctly fiscal products (though portions of them may eventually be employed for public purposes). They are simply the yields of the vineyard sections that Ioseph is overseeing.

The second and third texts do not stem from the Oxyrhynchite, but they merit inclusion for their unequivocality. Both are leases. The first of them, *P.Vind.Sal.* 9 (509), comes from the Hermopolite; in it, Sarapion, subdeacon (*hypodiakonos*) of the catholic church (*katholikē ekklēsia*) of Hermopolis, leases τὴν . . . μερίδα μίαν νεοφύτου (l. 6), "the one *meris* of the newly planted [vineyard]," that belongs to Aurelia Eucharistia. The other text, *SB* I 4482

48. As noted above (p. 44), there can be no doubt that *meris* (and not *meros*) is intended.

49. The quantities of wine involved, however, are small, too small to be a normal yield; they might represent some sort of charge.

50. Cf. also *P.Col.* VIII 239 (IV), an ostensibly private account noting the yields of two *lēnoi,* quantities of vessels that are further identified as "*from* [my emphasis] the *meris*" of a certain individual. For the meanings of *lēnos,* which was polysemic, see Dzierzbicka 2005, 27–50.

51. Alternatively, Ioannes and the *ampelourgoi* may have been partners; cf. chapter 3. One of these phenomena may recur in a text of unknown provenance, *P.Mich.* XV 744 (VI–VII). Here we encounter entries for the μερ( ) of various individuals; the associated quantities of wine are described as "through" other individuals. These quantities, however, are very small (at most 13 *dipla*), too small to represent yields.

52. Eds.: "It is my desire that the flow of wine should remain as it is for some time until the last *holding* [my emphasis] is ready, that we may not seem to press hardly upon the others who have not yet gathered the grapes."

53. It should be so unless the property in question was imperial.

(VI–VII), is from the Arsinoite. In this text, a party whose name is no longer extant rents property from someone who bears the title *lamprotatos;* the former individual acknowledges leasing "one *meris* of the vineyard that is surrounded by a clay [or "mud-brick"] embankment and a wall, the *meris* that belongs to it [antecedent: λαμπρότης] in the lands of the same hamlet in the place called 'Plamalo.'"[54] Both of these texts refer to concrete property, and both properties are explicitly viticultural.[55]

Finally, there is probably some archaeological evidence. As noted above, Grenfell and Hunt report that many of the Byzantine amphorae from Oxyrhynchus are labeled *chōriou* + geographical name, *mer( )* + personal name. It seems quite likely that *mer( )* should be resolved *mer(idos)* and that these notations indicate the precise origin of the wine.

Regarding Ownership

Owner, or Just Liturgist?

Gascou has written:

> The landowners of the Byzantine Egyptian cities formed a sort of college, sharing fiscal and liturgical responsibilities among themselves and representing all taxpayers of the jurisdiction before the municipal authorities. Thus, the men and products that the patrician Sophia and

---

54. τὴν διαφέρουσαν αὐτῇ ἐν πεδίῳ τοῦ αὐτοῦ ἐποικίου ἐν τόπῳ καλουμένῳ Πλαμαλῶ ἀμπέλου περιπεπλαστευμένης καὶ περιτετιχισμένης [read περιτετειχισμένης] μερίδα μίαν (ll. 3–6).

55. Cf. also *BGU* XVII 2682.9–10 (Hermopolis, 481), μερίδαν [read μερίδα] μίαν χωρίον ἀμπελικόν [read χωρίου ἀμπελικοῦ], "one *meris* of the viticultural *chōrion,*" as well as *P.Paramone* 18 (with Hagedorn and Mitthof 2004). For *meris* referencing nonviticultural concrete property during late antiquity, see *SB* XIV 11855 (Aphrodite, 547?).

Incidentally, *P.Vind.Sal.* 9 appears in Jördens' tabular overview of late sharecropping leases (*P.Heid.* V, pp. 232–38). Under her column labeled *Größe,* she records *Anteil,* but in all but one of the other texts with this entry, *meros,* not *meris,* appears. (See *P.Coll.Youtie* II 90 [Hermopolite, 513]; *P.Lond.* V 1841 [Aphrodite, 536]; *P.Panop.* 3 [310], 8 [338]; *SB* V 7666 [Panopolite, 330]. *SB* XIV 11855 [Aphrodite, 547?] has both *meros* [l. 14] and *meris* [l. 18].) It is striking that none of these other "Anteil" texts concern vineland, and one wonders if this does not account somehow for the distinction. Or perhaps the differentiation is between abstractly and concretely divided property? In an earlier (11.x.220) document, *P.Oxy.* LI 3638 (a "cession" of a share of a vineyard), *meros* is used to describe the object of sale. There the vineyard is said to be held in common by the seller and his siblings, and the buyer is the seller's half sister, so I suspect that an abstract share of the property is being sold, that there was no physical division of property. (This was often the modus operandi for house property.)

> Apion [e.g.] contributed as their share included not only their own contributions but also, and necessarily, those of other taxpayers for whom they were . . . held accountable.[56]

That the Apiones and other great magnates were responsible for the taxes of others seems uncontroversial; even Sarris, Gascou's most vehement critic, acknowledges that the estate collected sums from settlements for which it was "pagarchically responsible."[57] In itself, this is not problematic for questions of ownership; it creates challenges only because "public" and "private" are indiscriminately mixed and even inextricably meshed in the papers of landowners.[58] An Apion *pronoētēs,* for example, would collect rents and any taxes due from the tenants under his charge,[59] as well as the taxes of nontenants (and nonemployees) for whom the Apiones were liable under the share system (or Sarris's "pagarchy").[60] The latter sums might be identified in the *pronoētēs*' account,[61] but we must consult the rent roll for something approaching an itemized listing of charges, and even it does not indicate the Apiones' taxes on individual parcels of land.[62] Regarding this issue, Hardy has noted, "The Apion

56. Gascou 1985, 48.

57. Sarris 2006, 80. This is an incongruent admission in light of Sarris's estimate of the area of estate holdings. Cf., further, Ruffini 2010, 631–32.

58. This is not a novelty (nor should such blending be considered a hallmark of feudalism). For the blurring of "public" and "private" (or, rather, "Apollonian" and "personal") in the Ptolemaic Zenon archive, see Franko 1988, 71–73. Cf. also Meyer 2004, 30, on the accounts of magistrates in republican Rome.

59. Taxes due from employees, if not deducted from their wages, would also need to be collected.

60. Here it may be noted that *pronoētai* were not the sole collectors on the estate. *P.Oxy.* LXII 4350 (576), when considered alongside XVI 1917 (note especially ll. 120–21) and its structures, suggests that some *estate* properties were administered outside of the pronoesial system. (*P.Oxy.* LXII 4350 is relevant for the date of 1917, if the scribe Pamouthis reappears in 1917.120.) In *P.Oxy.* LXII 4351 (late VI), the inhabitants of the *ktēma* Pakerke agree to share the responsibility for the collection of the settlement's taxes with their *phrontistēs.* Pakerke is attested with a *pronoētēs,* however, in *P.Oxy.* XVI 1839 (VI) and (very probably) 2034.14 (VI), so 4351 must represent either a subcontract or a change in collection practice. For other fiscal collections by inhabitants of *epoikia,* see *P.Oxy.* XVI 1912.133 and XVIII 2195.97 (with Ast 2002, 181–82).

61. In, e.g., an entry tagged ὑπὲρ ἰδίας γῆς, "for his [*vel sim.*] own land" (cf. *P.Oxy.* LV 3804.92), which is no doubt short for ὑπὲρ δημοσίων ἰδίας γῆς, "for the taxes on his own land" (*PSI* VIII 954.23). Individuals with such property surely had other ties to their collector; cf. *P.Oxy.* XVI 1912.83–88 (before 566).

62. See *P.Oxy.* XVI 2037 (+ *BL* XI), a central text that has hardly received the attention that it merits. The unpublished versos of *PSI* I 76 contain similar accounts, but these presumably concern the holdings of Flavia Christodote (on whom see Keenan 1978 with *P.Thomas* 29 and *P.Oxy.* LXIX 4754). Editions of these three accounts will be published elsewhere.

accounts seem to indicate that payments to the government were made out of the *total* [my emphasis] receipts instead of being separated at the source as the money or grain came in from each tenant."[63] Perhaps more critically, an estate account that has recently come to light establishes that the *oikos* collected taxes from settlements in which it owned no or limited property.[64] We know only that these sites were villages (not their names), and they do not appear to have been subject to pronoetic administration. It is conceivable that they generated a documentary record that could suggest ownership when read out of context; in a text in which nonownership was the norm, it is far less likely that such status would have been indicated. In short, the accountants were not working with historians in mind.

It has been argued that the public responsibilities of landowners like the Apiones also included irrigation. Bonneau has written that "the administration of *sawāqī* ["water wheels," the Arabic for *mēchanai*], both for the distribution of water . . . and for the maintenance of the machines necessary for this distribution," was, "from the 5th century to the Arab period, semi-public and semi-private."[65] If she is correct—and I would suggest that documents like the massive axle distribution account *P.Oxy.* XIX 2244 indicate that she is—this has relevance for the identification of the owners of vineyards, since receipts for parts for irrigation machinery often describe a *mēchanē* as ἀντλοῦσαν εἰς ἄμπελον καὶ εἰς ἀρρώσιμον [read ἀρόσιμον] γῆν, "supplying water to vineyard and arable land."[66] In those receipts addressed to the Apiones (in which the *mēchanē* is, moreover, characterized as *geouchikē*, "of the landholder"), this would seem to indicate vineland belonging to the family. Some of the land referenced was surely owned by them: the machinery and the land attached to it were viewed as a unit, as is made clear by the regular (synecdochical) use of the term *mēchanē* to indicate a parcel of artificially irrigated

63. Hardy 1931, 51–52 (assuming, it seems, that tenants were responsible for the land tax; this was not typical in Egypt, though the landowner no doubt considered the tax when setting rents). Note also *CJ* XI.48.20 (529): *sin autem moris erat dominos totam summam accipere et ex ea partem quidem in publicas vertere functiones, partem autem in suos reditus habere* . . . , "If, however, it was the custom for the landlords to receive the total amount and from that to pay part for taxes, and to keep part as their return . . ." (in reference to taxes paid *through* a landlord, not directly). On the Apion estate, grain taxes were a district-level deduction (i.e., paid directly by the *pronoētai* from their total receipts), while net proceeds in coin were funneled to a higher level of management, whence money taxes were paid for the estate as a whole; see, most recently, Hickey 2008, 89–90.

64. See Hickey 2008, 89.

65. Bonneau 1993, 230.

66. I quote from *P.Oxy.* XVI 1990.18–19 (full ed.: *P.Oxy.* LXX, pp. 140–42; 591). That this is not simply a stock phrase is clear from other texts, where, e.g., only arable land is specified; cf., e.g., *P.Oxy.* XXXIV 2724.9 (469) and *PSI* I 60.15 (595).

land;[67] and it seems unlikely that the Apiones would have built or acquired the machinery without ownership of the land or sold the land under the machinery.[68] It is possible, however, that only a portion of the land referenced in the receipts belonged to the family, with the remainder of it simply receiving water as part of the Apiones' share of a relevant public service (*munus*).[69] The "pronoia clause" (καὶ εὐθέως ἡ ὑμῶν ὑπερφυεία πρόνοιαν ποιουμένη τῆς συστάσεως τῶν ἑαυτῆς πραγμάτων) does not indicate otherwise.[70] It has been translated, "And at once your magnificence, having regard for the state of your property,"[71] but *pronoia* here suggests curatorship,[72] *systasis* is not neutral but positive ("*good* state," "care," or similar), and *pragmata* ("concerns" or the like) should not be construed so narrowly that *munera* are excluded.[73]

Blurred Property Lines: Estate Land and Imperial Land

Viticulture on the Apion estate tends to be associated with smaller settlements (*epoikia/ktēmata*).[74] Generally speaking, the family had sole ownership of these entities (in contrast to the villages in which it had holdings),[75] but close

67. The two usages of *mēchanē* are clearly illustrated in a passage in the Coptic vita of St. Matthew the Poor (Amélineau 1895, 718):

> ⲁⲥϣⲱⲡⲉ ⲇⲉ ⲟⲛ ⲛⲟⲩϩⲟⲟⲩ ⲉⲣⲉ ⲟⲩϣⲏⲣⲉ ϣⲏⲙ ⲧⲁⲗⲏⲩ ⲉⲩϩⲟⲓ ⲉϥϫⲱⲣⲙ ⲛⲥⲁ ⲛⲉⲧⲉⲃⲛⲟⲟⲩⲉ ⲙⲡⲙⲱⲛⲁⲥⲧⲏⲣⲓⲟⲛ ⲙⲡⲉⲛⲉⲓⲱⲧ ⲁⲡⲁ ⲙⲁⲑⲉⲟⲥ ⲁ ⲡⲇⲓⲁⲃⲟⲗⲟⲥ ⲑⲓ ⲉⲡϣⲏⲣⲉ [*sic;* divide ⲑⲓⲉ ⲡϣⲏⲣⲉ] ⲕⲟⲩⲓ ⲉⲡⲉⲥⲏⲧ ⲉⲡϣⲏⲓ ⲁϥⲱⲙⲥ ϩⲙ ⲡⲙⲟⲟⲩ ⲁⲩⲱ ⲛⲧⲉⲣⲉ ⲟⲩⲛⲟϭ ⲛⲛⲁⲩ ϣⲱⲡⲉ ⲁϥⲱϣ ⲉⲃⲟⲗ ⲛϭⲓ ⲡⲙⲏⲏϣⲉ ⲧⲏⲣϥ ⲉⲧϣⲟⲟⲡ ϩⲙ ⲡϩⲟⲓ.
>
> And it also happened one day that a small boy was up on a *mēchanē* [i.e., *sāqīya*] goading the animals [i.e., the ones powering the *sāqīya*] of the monastery of our father Apa Matthew. The Devil made the little boy fall down into the *lakkos* [the *sāqīya*'s reservoir]. He sank in the water, and when a long time had passed, the whole crowd that was in the *mēchanē* [here clearly = "artificially irrigated parcel"] cried out.

68. Cf. also Bonneau 1993, 223, citing Constantinides 1969, 58: "It appears from the papyri that the irrigating machines belonged to the owner of the property upon which they stood."

69. Some sort of private distribution arrangement is also conceivable. For water rights under the Roman Empire, see Kehoe 2008 (an important contribution).

70. Here I cite *P.Oxy.* XXXVI 2779.13–15 (530). *P.Oxy.* XVI 1900 (528) is the first attestation of this form of the clause. Between 525 (*P.Oxy.* LXX 4781) and 530 (4785), we find φροντίδα in place of πρόνοιαν, while a somewhat sparer version (φροντίδα τιθέμενον τῶν αὐτοῦ πραγμάτων, "having care for its concerns") appears already in *P.Oxy.* XVI 1982 (497).

71. *P.Oxy.* XXXVI 2779, p. 75.

72. As the use of φροντίς in its place suggests; cf. *P.Oxy.* LXX 4781, 13n.

73. Cf. Bonneau 1993, 228–29.

74. Cf. appendix B, where these estate properties outnumber *kōmai* by approximately 6:1; about half of the references to villages, moreover, concern purchase (or exaction) of wine and are not related to viticultural activity on Apion property.

75. Cf., e.g., Jones 1964, 787 and n. 41.

examination of texts in the estate dossier reveals a more fluid and varied situation. *P.Oxy.* XVI 1915, which dates to shortly after 557,[76] well illustrates the blurring of property lines between imperial holdings of this type and similar properties belonging to the Apiones. This account concerns sums due from the κτῆμα Πεμπὼ τοῦ θειοτάτου οἴκου (l. 1). Although it is imperial property, Pempo's *lēmmata* ("inflows" or "receipts") are said to be contained in the *apaitēsimon*, or rent roll, of the Apiones' glorious house.[77] It is probable that the imperial lands at Pempo (4 *arourai* of *ampelos* and 61.5 of grain land) were being administered by the Apiones and, moreover, that this was the state of affairs for a long period;[78] compare the entries tagged καθὼς περιέχει τὸ ἀπαιτήσιμον ἐξ ἀρχαιωθέντος, "as contained in the rent roll from time immemorial."[79] Some of the land at Pempo was already being charged to (and presumably being cultivated by) individuals from the neighboring *ktēma* Meskanouneos and the *epoikia* Kineas and Megalou Choriou (ll. 10–13), and in lines 16–20, an additional 20 *arourai* have been assigned to farmers (*geōrgoi*) from the first two of these sites (and probably from the last of them as well). All three of the settlements receiving the transfers are known from other texts to have belonged to the Apiones, but the *epoikion* Kineas is also attested as part of the *domus divina*.[80] The assignment of land to other settlements—here indicated by the verb *rhiptō* (literally, "throw")—is obviously reminiscent of an earlier phenomenon known as *epimerismos* (the compulsory cultivation

76. For this date, cf. Mazza 2001, 24 n. 73.

77. Accordingly, they likely appeared among the *lēmmata* in the accounts of the Apiones' *pronoētai*, since these were derived from the *apaitēsimon;* cf. Mazza 1998, 171. Given accounting practice (cf. the previous section of this chapter), one wonders how frequently the *pronoētai* identified them as imperial.

78. Johnson and West (1949, 74) propose another alternative for 1915: "Quite possibly this property represents an emphyteutic grant of imperial land to the house of Apion." Presumably this interpretation draws on phrases like ἐξ ἀρχαιωθέντος καὶ ἀμνημονεύτου χρόνου ("from time immemorial," l. 5) and the fact that the imperial property appeared in the Apiones' *apaitēsimon.* Although Johnson and West's scenario is possible, the references to established practice indicate only that, and we have already seen that the accounts of the *pronoētai* (and therefore the *apaitēsimon;* cf. the preceding note) could contain property that was not being exploited by the estate. Also note the discussion of *apotakton chōriōn* in the next section of this chapter.

79. I quote line 7; ἐξ ἀρχαιωθέντος is shorthand for ἐξ ἀρχαιωθέντος καὶ ἀμνημονεύτου χρόνου (cf. the preceding note). It is surely not a coincidence that the first securely identifiable member of the Apion family, Strategius I, was involved with the administration of imperial estates.

80. For the *ktēma* Meskanouneos and *epoikion* Megalou Choriou, see *P.Oxy.* LXXII 4926 (564), 2n., 3n., and refs. (*P.Mert.* II 96 probably reflects the situation attested by *P.Oxy.* XVI 1915.) The *epoikion* Kineas belongs to the *domus divina* in *P.Oxy.* LXXII 4901 (408) and *PSI* III 196, 197 (HGv: VI); presumably the latter two texts predate XVI 1915 (given the ownership of Meskanouneos and Megalou Choriou in 1915). In *SB* XXII 15362 (614/15), Kineas belongs to the Apiones; *P.Oxy.* XXVII 2479 (HGv: VI) also indicates ownership by the estate.

of abandoned or ownerless land),[81] and a comparable practice, *epibolē,* was in use during the period under consideration here.[82] There is no way to know the extent of such transfers elsewhere on the estate or whether the latest transfers documented by 1915 were permanent. The transfers in 1915 seem to have been compulsory (l. 16), but others may not have been.[83]

ἀπότακτον χωρίων

In *P.Oxy.* 1915, a sum due for the 4 *arourai* of imperial vineland is described as ὑπὲρ ἀποτάκτου χωρίων.[84] This designation also appears with some frequency in the accounts of the Apiones' *pronoētai,* and it is relevant for the issue of ownership because Gascou has identified it with emphyteusis.[85] If he is correct, the accounts of the *pronoētai* contain property under emphyteutic lease, that is, land that probably belonged to the state or the church and that the Apiones were exploiting themselves or subletting to other individuals. But is Gascou's assertion valid? It seems to hinge solely on a long-term lease from the Antinoopolite nome (*P.Cair.Masp.* III 67299), where the phrase ὑπὲρ ἐτησίου ἀποτάκτου ἤτοι ἐμφυτεύματος occurs (ll. 39–40).[86] The word *ētoi,* however, can be construed in several ways. I presume that Gascou interprets it as meaning "also known as."[87] I myself would suggest an explanatory function—in other words, that one should understand here that an emphyteutic

81. Cf. Maresch 1994, 83 and n. 6; the standard treatment of *epimerismos* does not provide any examples from after the fourth century (see Poethke 1969, 63–66). Gagos and Sijpesteijn (1995, 245–47) report that *epimerismos* was replaced by *epinemēsis* under Diocletian, but note the criticisms in Bagnall 2003. Further study of these and related practices would be welcome.

82. Cf. Jones 1964, 813–15. The Latin term for the practice is *adiectio sterilium;* the Greek *epibolē* first appears in *CJ* I.34.2 (Anastasius). For its relevancy to imperial land, cf., e.g., *CJ* XI.59.9 (394).

83. For the issue of compulsion, cf. Jones 1964, 814–15. Sarris (2006, 83) has written, "The Apion estates expanded, the papyri reveal, at the expense of both private landowners and the imperial household." This is an accurate statement, but the relationship with the *domus divina* was surely more nuanced. *P.Oxy.* XVI 2055 (VI) reveals the flight of Apion *geōrgoi* to the *domus divina*—were they enticed? Note also *P.Oxy.* XVI 2030 (VI), with the discussion of *dikaion* below; in this text, Pempo seems to be paying for a portion of the *epoikion* Meskanouneos. Needless to say, one's reading of the texts is dependent on one's conception of the relationship between the state and large landowners during this period; cf. Keenan 1993, 142 and n. 31. Banaji's comments on this issue (2000, 96–98) very much merit our attention.

84. Line 6. χωρ( ) is incorrectly resolved as χωρίου throughout the editio princeps and, e.g., in Maresch 1994, 82–90.

85. Gascou 1985, 7–8 and n. 16 (not revised in Gascou 2008, but there 1985 is presented "nearly in its original form" for reasons explained on p. 125). Cf. already Comfort 1937, 10 n. 3.

86. *Apotaktos: M.Chr.* 134, 15n., e.g., defines φόρος ἀπότακτος as "fixed rent, in contrast to sharecropping."

87. Cf. P. J. Sijpesteijn 1992, 242.

rent is a specific kind of fixed annual rent.[88] In any case, it is abundantly clear from the papyri of the Byzantine period that not every rent that is described as "fixed" is emphyteutic.[89] This is important to remember, since the payments in the Apion accounts make specific reference to *apotakton chōriōn,* and the second element in this expression is neither explicit nor implied in the Maspero papyrus.

What, then, is *apotakton chōriōn*? Maresch, in his discussion of *P.Oxy.* 1915, translates it as "the assigned land";[90] he does not comment further. Perhaps he has been led to this conclusion by the echoes of *epimerismos* in 1915; or maybe he has been influenced by Johnson and West, who thought that ἀπότακτα χώρια (*sic*) and ἐκτάκτη τρύγη (*sic*) were "apparently arable land and vineyards which lacked tenants, or an epibole."[91] Banaji once argued that the term described land whose rent was fixed because it was irrigated artificially.[92] *Apotakton* doubtless does refer to a fixed rent, and the association of perennial irrigation with such rents is plausible, but it is possible to be more precise.

Table 2.1 collects all of the references to *apotakton chōriōn* from the papyrological corpus. Every attestation involves the Apiones. The designation *may* refer to land owned by the Apiones, but in two cases (2.1.7–8), it is almost certainly used for imperial land that the family is administering. Most remarkable are the links to viticulture. It has already been established that the translation "vineyard" should be our *point de départ* when we encounter *chōrion* in the papyri of the Byzantine Oxyrhynchite. In 2.1.7, the land subject to *apotakton chōriōn* is explicitly vineland, while in 2.1.3–4, 13–15, 19, 21, and probably 10, the parties include *ampelourgoi* (vineyard specialists). Other attestations are suggestive. In 2.1.1, 17, and 18, for example, the places concerned are otherwise known to have had vineyards; in 2.1.6, the land seems to be associated with *neophyta chōria.* Similarly, the account entry immediately preceding 2.1.9 concerns supplementary dues connected to vineland.[93] Table 2.1.16 has an explicit reference to wine (though *chōriōn* does not appear). Finally, the charge of 3

88. Cf. P. J. Sijpesteijn, 1992, 245–46. That *ētoi* here approaches *kai* (cf. Sijpesteijn, p. 241 n. 7)—i.e., that the rent is being described (redundantly) as both fixed and emphyteutic—seems unlikely. Banaji (2007, 95) interprets *ētoi* as I do.

89. To my knowledge, the only explicit reference in the late Oxyrhynchite documentation to an emphyteutic lease is *P.Oxy.* LV 3805.12. Cf. also *P.Oxy.* LV 3803 + *BL* IX 204 and Gonis 2000a, 95.

90. Maresch 1994, 87.

91. Johnson and West 1949, 64.

92. Banaji 1992, 225–28; this assertion is absent, however, from 2007, 94–96 (where only his [valid] criticisms of Gascou appear).

93. Cf. also 2.1.19. For these *ektakta trygēs,* see p. 70 below.

TABLE 2.1. *Apotakton Chōriōn*

Reference	Party/Sum	Notes
1. *P.Oxy.* VI 998.5[a]	οἱ ἀπὸ Νεκώνθεως/22 *solidi,* Alexandrian standard,[b] + 12 *keratia* = 22 *sol.,* 8.75 *ker.* on the public standard	—property administered or owned by Apiones —a concession (*parachōrēsis*) —vineyards documented here[c]
2. *P.Oxy.* XVI 1912.71, cf. *BL* XI 155	οἱ ἀπὸ Οὐάλεντος/3 *sol.*	—property administered or owned by Apiones
3. *P.Oxy.* XVI 1912.81	οἱ αὐτοὶ γεωργοὶ καὶ ἀμπελουργοί, ?/18 *sol.*	—property administered or owned by Apiones
4. *P.Oxy.* XVI 1912.88, cf. *BL* XI 155	οἱ αὐτοὶ γεωργοὶ καὶ ἀμπελουργοί, *epoikion* Samakionos/8 *sol.*	—property administered or owned by Apiones
5. *P.Oxy.* XVI 1912.136	οἱ ἀπὸ Οὐάλεντος/?	—property administered or owned by Apiones —a concession
6. *P.Oxy.* XVI 1912.137–39	Thaesis from *epoikion* Chenetorios; text very uncertain but land probably in *epoikion* Kleonos, *mēchanē* Psasenne[d] /.373 *sol.* for 1 *ar.*	—property administered or owned by Apiones —a partial concession (for a *phoros*) —land described as *enchersos* ("dry") —verb *trygaō* ("harvest," often in reference to grapes) connected with land —in l. 153, this *mēchanē* receives new reeds for 1 *ar.* of newly planted vineland
7. *P.Oxy.* XVI 1915.6	*ktēma* Pempo, 4 *ar.* vineland @ 12 *sol.* on the public standard	—imperial property almost certainly administered by Apiones
8. *P.Oxy.* XVI 1915.21	*ktēma* Pempo, 30 *ar.* @ 10 *sol.*	—imperial property almost certainly administered by Apiones —entry canceled in text (an error)
9. *P.Oxy.* XVIII 2195.18	τὸ κοινὸν τῶν γεωργῶν, *epoikion* Tillonos?[e]/?	—property administered or owned by Apiones —immediately preceded by a payment from same for *ektakta trygēs*

(continues)

TABLE 2.1—*(Continued)*

Reference	Party/Sum	Notes
10. *P.Oxy.* XVIII 2195.71	τὸ κοινὸν τῶν γεωργῶν καὶ ἀμπελουργῶν?,[f] *epoikion* Erotos/9 *sol.*	—property administered or owned by Apiones
11. *P.Oxy.* XVIII 2195.124	οἱ ἀπὸ Τίλλωνος, *mēchanē* Kollothou/1 *sol.*	—property administered or owned by Apiones —a concession
12. *P.Oxy.* XIX 2243a.29	οἱ αὐτοὶ καὶ προκείμενοι γεωργοί from *epoikion* Pesta/7.5 *sol.* without *rhopē*	—property administered or owned by Apiones
13. *P.Oxy.* LV 3804.34	τὸ κοινὸν τῶν γεωργῶν καὶ ἀμπελουργῶν, *epoikion* Apele/43.5 *sol.*	—property administered or owned by Apiones
14. *P.Oxy.* LV 3804.47	τὸ κοινὸν τῶν γεωργῶν καὶ ἀμπελουργῶν, *epoikion* Pakiak/13.5 *sol.*	—property administered or owned by Apiones
15. *P.Oxy.* LV 3804.101	τὸ κοινὸν τῶν γεωργῶν καὶ ἀμπελουργῶν, *epoikion* Loukiou/9 *sol.*	—property administered or owned by Apiones —Note the identical entry in *SB* XXIV 16324.31.
16. *P.Oxy.* LV 3805.68–69	οἱ κληρονόμοι Ἱερημίου μείζονος Πακέρκυ/9 *sol.*, 1.5 *ker.*, on the Alexandrian standard = 9 *sol.* on the public standard	—document from Apion dossier —only ἀποτάκτου appears; οἴνου superscript
17. *P.Oxy.* LV 3805.76–78	οἱ ἀπὸ Νεκώνθεως/22 *sol.*, 15.75 *ker.*, on the Alexandrian standard = 22.5 *sol.* on the public standard	—document from Apion dossier —vineyards documented here[c]
18. *P.Oxy.* LV 3805.87–88	οἱ ἀπὸ Ταρουσὲβτ καὶ Λουκίου τῶν κτημάτων/10 *sol.*, 16.75 *ker.*, on the Alexandrian standard = 11.25 *sol.* on the public standard[g]	—document from Apion dossier —*ktēma* Tarouseb(t) known to have vineyards[h]
19. *PSI* VIII 954.22	τὸ κοινὸν τῶν γεωργῶν καὶ ἀμπελουργῶν, *epoikia* Skytalitidos and Krariou/52.5 *sol.*	—property administered or owned by Apiones —*ektakta trygēs* from heirs of *geōrgoi* of *epoikion* Skytalitidos, l. 7.

TABLE 2.1. *(Continued)*

Reference	Party/Sum	Notes
20. *PSI* VIII 954.29	οἱ αὐτοὶ καὶ προκείμενοι γεωργοί, *triton meros* of *epoikion* Chrysochoou?/ 27 *sol.*	—property administered or owned by Apiones
21. *PSI* VIII 954.41	τὸ κοινὸν τῶν γεωργῶν καὶ ἀμπελουργῶν, *epoikion* Psempekle/9 *sol.*	—property administered or owned by Apiones —*apotaktou* written απο( ) τακτου
22. *PSI* VIII 954.48	τὸ κοινὸν τῶν γεωργῶν, *epoikion* Ptolemaidos/?	—property administered or owned by Apiones

Note: Also observe *SB* XX 15097, which almost surely comes from the Oxyrhynchite (given the presence of a *mēchanourgos*). Unfortunately its editor neither provides a discussion of the relevant lines (ll. 4–5), nor does he offer his interpretation of the text. Moreover, there is no photo against which one might check the readings.

In addition, see Ast 2002, 185–86, 5f.n., on *PSI* VIII 954.42.

[a]The text is a *descriptum* and will be published by the author and Amin Benaissa.

[b]For this and related monetary terminology in this column, cf. *P.Oxy.* LXXII 4390, 13–16n.

[c]See appendix B, s. ΝΕΚΩΝΘΕΩΣ.

[d]Cf. remains of text with l. 153.

[e]Cf. appendix B, s. ΤΙΛΛΩΝΟΣ.

[f]Cf. Ast 2002, 185, n. 25.

[g]Line 87 deducts 105 *keratia* from the "public" *solidi;* this is an error for 90 (which I have used in my calculation here).

[h]See appendix B, s. ΤΑΡΟΥΣΕΒ(Τ).

solidi per *aroura* attested in 2.1.7,[94] as well as the generally large sums of *solidi* involved (note especially 2.1.13 and 19), suggests that we are dealing with rather valuable property like vineland. It comes as no surprise that the parties paying *apotakton chōriōn* are, in the great majority of cases, collectives—the people from such-and-such settlement, the community (*koinon*) from such-and-such, and the heirs of so-and-so.[95]

In light of this evidence, it seems rather probable that the Apiones used *apotakton chōriōn* as a designation for a fixed charge—or, better, rent (*phoros;* cf. 2.1.6)—on vineyards. There are indications, however, that even more precision is possible, specifically that the term was used *only* for imperial vineyard property that was being administered by the family (note again 2.1.7–8). *P.Oxy.*

94. Such a rate, incidentally, is not what one would expect under emphyteusis; cf. Banaji 2007, 95.

95. "Heirs of so-and-so": Collectivity may not be the decisive factor in this case (2.1.16), but rather the fact that the deceased was a *meizōn*. For the wealth of these village headmen, cf., e.g., *P.Oxy.* I 136 (an estate of 360 *sol.* divided between three heirs) and XVI 2058 (*stolen* property valued at 86 *sol.* on the private standard).

LV 3804.135, for example, records a large sum (13 *sol.*) owed by a *koinon* of *geōrgoi* and vineyard specialists for the "ground" (*edaphos*) of the *ampelourgoi;* though *apotakton chōriōn* is not mentioned, the most plausible interpretation is that this represents a rent for vineland.[96] This raises the question of what exactly the distinction was between *apotakton chōriōn* and the arrangement governing the vineyard in 3804.135. I would suggest that it was one of ownership. It is striking that *apotakton chōriōn* does not account for a higher percentage of the receipts in some of the *epoikia* at which it is attested; at the *epoikion* Erotos, for example, in the north of the nome where flax is not likely to have been a source of income,[97] *apotakton chōriōn* yields only 24 percent of the total *lēmmata* in cash. It seems most economical to attribute the other 76 percent to Apion-owned vineyards (and other estate properties that yielded cash rents). The charges for *apotakton chōriōn* themselves support this hypothesis. With one exception (2.1.1), all are neatly divisible by three *solidi,* the rate, on the public standard, that was charged the imperial property in 2.1.7–8. If we convert this public standard rate to the private standard,[98] the standard on which estate rents and charges would generally have been calculated, quotients that include irregular fractional *arourai* result.[99] In short, although he was incorrect regarding emphyteusis—the Apiones were probably administrators—Gascou seems to have been justified in linking *apotakton chōriōn* to institutional land.

τὸ δίκαιον

In estate documentation, particularly in accounts, one often encounters the term *dikaion.* Ruffini renders this as "foundation," without further comment; he apparently considers it a type of settlement.[100] Hardy suggests that it refers

96. Note also, e.g., *PSI* VIII 954.42 with *P.Mich.* XV 734 (for ι̣ε̣είου in l. 10 of this text, read Ψεείου): a very high cash rent (27 *sol.*) at a site with known wine production.

97. See p. 34 above.

98. Whether we interpret the gold standards like Banaji (1998, 189: 3 *sol.* public = 3.333 *sol.* private) or Maresch (1994, 43: 3 *sol.* public = 2.7 *sol.* private).

99. This assumes, of course, that the figures in table 2.1 that have been drawn from the *lēmmata* of pronoetic accounts represent sums booked (or weighed) on the public standard. While the numismatic phenomena attested in estate documentation are beyond the scope of the present study (and, indeed, merit a monograph in their own right), I would suggest, as a working hypothesis, that *all* imperial rents (not just charges for *apotakton chōriōn*) were booked in the pronoetic accounts on the public standard. In other words, payments from imperial land may account, at least in part, for unexpectedly high carat deductions (> 5 *keratia/sol.*; cf. the data in Maresch 1994, 126–28) in transfers from *pronoētai* to Apion bankers.

100. Cf. Ruffini 2008, 108.

to lands "acquired from former owners" that were kept separate in the accounts for the "purpose of avoiding changes."[101] Johnson and West have a bit more to say:

> Somewhat more perplexing is the use of δίκαιον in the accounts. Sometimes it seems to describe the bailiwick of a particular pronoetes, or even the particular divisions or ἐποίκια of a prostasia, but the δίκαιον τοῖς τεταρτομοιρίταις [*sic*], the δίκαιον of Diogenes, and of Cephalas seem to be different. The last of these seems to have been administered by a special pronoetes and seemingly paid no tax. This may have been a special grant of the Emperor like the dorea of Protamus at Hermopolis. Still more obscure is the δίκαιον ἀντικαταλλαγῆς τοῦ δεῖνος found at Heracleopolis. The phrase is not found elsewhere. It may imply that the owner of this estate collected the rents for some of his friends who may have owned small allotments contiguous to his estate. Grenfell and Hunt suggest that Apion may have collected revenues from imperial estates.[102]

A reading of the texts suggests that Hardy is closest to the mark. There are a few grains of truth in Johnson and West's more complete discussion, but there is also much confusion—perhaps justifiably, given the ambiguity that *dikaion* (like the Latin *ius*) can have. *P.Oxy.* LXIII 4397 (545) confirms Hardy's assertion about former owners, at least with regard to the *dikaion* Diogenous. This rich text recounts how Strategius II had come into possession of Diogenes' holdings when the latter died with outstanding loans owing to the former. Diogenes' brother and sister understandably had refused their inheritance, and as a result, they forfeited the property to Strategius. All of this happened before 542, but references to the *dikaion* Diogenous still appear in, for example, 577.[103] "Besitztum, Besitzgegenstand" seems to be the best rendering in the lexica for the usage of *dikaion* with which we are concerned,[104] though it is remarkable that the designation would still appear in texts thirty-five years after the transfer transaction.[105] The best explanation is a fiscal one:

101. Hardy 1931, 101.

102. Johnson and West 1949, 64.

103. *P.Oxy.* XVIII 2195. Incidentally, in *PSI* VIII 954.8, Διογένους is immediately preceded by δικαι( ), allowing us to assign this text (currently dated "VI") to the period after the Diogenes incident.

104. *WB*, s.v. δίκαιον, n. 4.

105. It appears, however, only in the (fixed/fossilized; cf. Mazza 1998, 166–67) *lēmmata* of *P.Oxy.* XVIII 2195. We encounter it in the (annually variable) *analōmata* of LV 3804 (565/66) at line 203. For the locations of Diogenes' property, see Mazza 1998, 135 n. 246 (and note that Pleein and Plakitou may be localized to the vicinity of the seventh *pagus*).

the specification presumably served as an aide-mémoire or cross-reference for those needing to consult unrevised fiscal registers (the *dēmosioi chartai/ptyktoi*).[106] No *dēmosiai apographai* (Latin *census*) are known from the Byzantine Oxyrhynchite, but the Aphroditan evidence is instructive.[107] There, the *summetrēsis* of the *censitor* and *scholastikos* Ioannes (*SB* XX 14669; ca. 523) was still being referenced ca. 566.[108] Moreover, a convention in this *summetrēsis* and similar documents, that of entries with P(ersonal) N(ame) ὀνόματος PN, which indicates land possessed by one person but registered under another name,[109] is suggestive of the usage just proposed for *dikaion* in the Apion accounts.[110] If this interpretation is correct, *dikaion* entries provide some insight into the expansion of the estate, including its acquisition of vineyards. The vineyards in Text I, for example, probably belonged to the *dikaion* Kephala; that is, they were the former possessions of an unknown Kephalas.[111] Another attested *dikaion*, that of Theodoulos, may well indicate former land of a known *geouchos*.[112]

My explanation is also applicable to those usages of *dikaion* in which no personal name follows. Johnson and West seem to have divided these into two groups: those that reference "the bailiwick of a particular pronoetes" and those that define "the particular divisions or ἐποίκια of a prostasia." *P.Oxy.* XVI 1911.109–10, which Johnson and West cite under the first definition, reads,

106. *Dēmosioi chartai:* cf. *P.Warr.* 3.9 (Oxy., < 504). *Ptyktoi: P.Oxy.* I 126.8 (572). The change in terminology (papyrus to codex) may well be significant (and the restorations of χαρτῶν in *P.Oxy.* XVI 1886.6 [538] and *SB* XXIV 15955.6 [Oxy., 540 or 541] are unjustified).

107. Fiscal documentation from the Byzantine Hermopolite is also helpful; cf. *P.Sorb.* II 69, pp. 20–32 (e.g., p. 21: "A number of our *onomata* had nothing more than a purely fossil existence when the codex was drafted").

108. *P.Michael.* 42A.19–21. Note also Gascou and MacCoull 1987, 105 n. 6. Incidentally, this *summetrēsis* was preserved in a codex (cf. n. 106 above). Zuckerman (2004, 36) defines a *summetrēsis* as a "supplementary survey."

109. According to Zuckerman (2004, 29), *onoma* designates "the holder of a cadastral entry."

110. On the Apion estate, note the *dikaion* Kephala and *P.Oxy.* XVI 1916r.1, † λόγος τοῦ ὀνόματος Κεφαλᾶ. Fourth-century legislation indicated the need for prompt registration of a new owner (*CTh* XI.3.5), but this was not incorporated into the *Codex Justinianus,* which emphasizes the payment of the tax regardless of registration (e.g., *CJ* IV.47.3 = *CTh* XI.3.3.). A change of registration was accomplished by an *epistalma tou sōmatismou.* Note also the term *ekbolē* in Ioannes' cadastre (with Gascou and MacCoull 1987, 109–10). (Banaji 2007, 195, has another interpretation.)

111. Johnson and West's assertion (in the block quote above) that the *dikaion* Kephala was "administered by a special pronoetes and seemingly paid no tax" has little basis in the texts. *P.Oxy.* XVI 1916r (ca. 547) suggests that the holdings of Kephalas were incorporated into as many as seven *prostasiai,* including that of Kotyleeiou, an *epoikion* for which we also have a *pronoētēs*' account. Curiously, the *dikaion* is not mentioned in the relevant *lēmmata* (or in the *analōmata*) of this account (cf. *P.Oxy.* LV 3804 [566]). Or are we to see the *dikaion* Kephala in δικαίου τῶν υἱῶν τοῦ ἐπισκόπου (*P.Oxy.* LV 3804.127; note l. 129 of this text and *P.Oxy.* XVI 2020.15)?

112. *P.Oxy.* LV 3804.100 and *SB* XXIV 16324.30. For Theodoulos, see Gonis 2002b, 91.

ἐδάφους Φηλτανβὲλ γεωργουμένου ποτὲ παρὰ Φιλέου Ψαειᾶ ἀπὸ Λουκίου δικαίου τοῦ κτήματος Ταρουσέβ, "ground of Pheltanbel, formerly farmed by Phileas, son of Psaeias, from Loukiou, right of the holding Tarouseb";[113] this merely indicates that Phileas' *origo* was in a part of the *ktēma* Tarouseb that had come to be administered in the *ktēma* Loukiou.[114] Tarouseb is simply a cross-reference; both of the settlements mentioned were in the "bailiwick" of a single *pronoētēs.*[115] Johnson and West cite 1911.107 as an example of their second definition, but there ἐν αὐτουργίᾳ δικαίου τοῦ κτήματος Ταρουσέβτ simply means "in the directly worked land that (on the cadastre) is part of the holding Tarousebt." Finally, there are the δίκαιον ἀντικαταλλαγῆς τοῦ δεῖνος and the δίκαιον τεταρτομοιρίας (and similar formulations). The confusion of Johnson and West concerning the former is puzzling, for it no doubt simply refers to property acquired (from ὁ δεῖνα, "so-and-so") by barter (*permutatio*), as opposed to sale (*prasis*).[116] The latter designates property that had come to its possessor through the municipal council, which itself had acquired said property as a result of the laws governing curial succession.[117]

---

113. I quote the text as corrected in *BL* IX 190.

114. Cf. *P.Oxy.* XIX 2244, p. 142, 6n.: "The sense appears to be that the μηχανή *for administrative purposes* [my emphasis] was included in the estate named."

115. Cf. *P.Oxy.* LV 3804, p. 96.

116. For the Apion references, see *P.Oxy.* LV 3805, p. 170, 65–66n. For the contract type (*antikatallagē*), cf., e.g., *P.Kellis* I 30, which also makes reference to a property acquired ἀπὸ δικαίου πράσεως (l. 14), "by right of sale." Note also *P.Lond.* V 1769.5, a *meris* of a *chōrion* ἐλθὸν εἰς σὲ ἀπὸ δικαίου ἀγορασίας παρὰ Ἰσακίου Σαραπίωνος, "that came to you by right of purchase from Isakis, the son of Sarapion."

117. Cf., e.g., *CJ* X.35.1 (428). For discussion, see Fikhman 1974 (overstated) with *P.Oxy.* LXVIII 4687, 8–9n.

Chapter 3

Labor in the Vineyard

The State of the Question

"[In Egypt] wage labour was widely used, in particular but not only on the Apion estates," Wickham has declared.[1] If he is correct, we would expect that direct exploitation was most pronounced in those agricultural sectors that required the greatest investments of resources and manpower, including (even especially) viticulture.[2] Wickham has chiefly relied on the research of Sarris and Banaji for his discussion of the Apion estate. Sarris's principal support for direct exploitation in estate viticulture seems to be *P.Oxy.* XVI 1859, a letter in which a *chartularius* begins (ll. 1–4), "I exhort my good master, if it is possible, to order that Pambechios and his companions from Seryphis be spared today and tomorrow, until they complete their landlordly yield [of wine], since today they began to press [the grapes of] their vineyard. For they are producing a great yield."[3] There is nothing here that unequivocally indicates wage labor.

1. Wickham 2005, 274.

2. Cf. Bagnall 1993b, 134.

3. παρακαλῶ τὸν ἐμὸν ἀγαθὸν δεσπότην ἐάν ἐστι δυνατὸν κελεῦσαι ἐνδοῦναι τὸν Παμβήχιον καὶ τοὺς ἑτέρους (read ἑταίρους) αὐτοῦ ἀπὸ Σερύφεως τὴν σήμερον καὶ τὴν αὔριον ἕως οὗ πληρώσουσιν τὴν γεουχικὴν ῥύσιν αὐτῶν, ἐπειδὴ σήμερον ἤρξαντο τρυγεῖν τὴν ἄμπελον αὐτῶν. πολλὴν γὰρ ῥύσιν παρέχουσιν. Cited by Sarris (2006, 74), who renders ἕως οὗ πληρώσουσιν τὴν γεουχικὴν ῥύσιν αὐτῶν as "until they finish their share [*sic*] of the estate wine yield." Note also his p. 177 n. 22 (comparative evidence from Merovingian and Carolingian Gaul), elaborated in Sarris 2004a, 309–10. I owe the rendering of τρυγεῖν as "press" (as opposed to "harvest") to a suggestion from one of this volume's anonymous referees.

Pambechios and his partners—who are villagers, not *epoikiōtai*[4]—could certainly be tenants, and the *chartularius*' request might be coming in response to a demand for labor service elsewhere.[5] The "landlordly yield" could well be a rent in kind.[6] That these men were being supervised is not determinative; the estate would have been equally interested in ensuring that it received the rents that it was due. Banaji, for his part, is silent about labor in the Apiones' vineyards in his 2007 monograph,[7] but an earlier article does address the question.[8] Here Banaji informs us that "wage labour was used extensively" on the aristocratic estates of the sixth century, and he cites six texts in support of this statement.[9] Of the papyri that pertain to the Apiones, only one, *SPP* III[2] 86, concerns work in vineyards; in this receipt, a *riparius* of the *endoxos oikos* in the Herakleopolite, who seems to be functioning in a capacity akin to that of a *pronoētēs,*[10] makes a payment to a *geōrgos* "for the wage for the irrigation work which I [the *geōrgos*] am doing in the vineyards under your reverence [referring to the *riparius*] . . . in the *meris* of the *ampelourgos* Paomis, the son of Kelebes."[11] Banaji also argues (210–11), based on several *prochreia* (cash advance) agreements made with *geōrgoi* associated with estate *mēchanai* and other texts, that "*georgoi* or other agricultural labourers such as those tending gardens or vineyards were assigned to individual irrigated plots or farms (or gardens or vineyards), or vice versa, usually paid cash wages (in the prevalent gold currency), and entitled to advances out of them [the *prochreia* agreements]." Yet while it is certainly plausible to associate *mēchanai* with vineyards, which could not survive without perennial irrigation, the texts that

4. *Epoikiōtai* are dwellers of estate-owned *epoikia,* who are more likely to have been employees; cf. below (pp. 85–86).

5. I am thinking of the *geouchikai hypēresiai* in leases, for which see Fikhman 2006, 228–30; note also Mazza 2001, 161–62. This is but one of several possibilities. Pace Sarris 2006, 64–65, the *hypēresiai* in *P.Oxy.* XXVII 2478 are no doubt connected to Pambechios' position as tenant, not his status of *enapographos.*

6. As it probably is in *P.Oxy.* XVI 1896, also cited by Sarris. On this text, see below (p. 65).

7. An update of the 2002 edition. This volume was based on his dissertation, which did include relevant arguments (Banaji 1992, 224–25). Since the content in question seems not to have been incorporated into any of his publications on the economy of the later Roman Empire, I am not inclined to argue against it.

8. Banaji 1999.

9. Banaji 1999, 205 and n. 78: "*SPP* III 86; *P.Oxy.* XVI 1911.157–59 (with *BL* 9.191), 1913.16–18, 19, 21–23, XVIII 2195.134–36; *SB* VI 9284 (553); *P.Grenf.* I 58 (561)." The last of the texts cited involves more than a simple wage; for discussion, see *P.Heid.* V, p. 266.

10. For the usual duties of *riparii,* cf. Torallas Tovar 2001.

11. Lines 3–4: ὑπὲρ μισθοῦ ἀρδείας ἧς ποιοῦμαι τῶν ὑπὸ τὴν ὑμετέραν αἰδεσιμότητα ἀμπελικῶν χωρίων . . . μερίδος Παωμίου Κελεβης ἀμπελουργοῦ. It is also conceivable that *hypo* indicates that the *riparius* was a tenant. The *ampelourgos* Paomis either was a tenant (or subtenant, if the *riparius* was a tenant) or had this vineyard section assigned to him.

Banaji cites make no mention of viticulture or the work associated with it; their only sure connection is to the supply of water.[12]

All in all, the data underlying recent claims for the direct exploitation of the Apiones' viticultural land are unsatisfying. To some degree, this is a consequence of the poverty of the evidence: not a single vineyard lease from the estate is extant, for example, and we must rely mostly on accounts and receipts for our information; the proper interpretation of this data, moreover, is not always clear. There is, however, more evidence than has hitherto been considered, in addition to a wealth of unedited documentation. Yet more troubling has been the tendency to schematize the estate's deployment of labor, or, to put it another way, to privilege an overarching theory over evidence. To be sure, the rich data set furnished by the papyri is of limited usefulness without some sort of interpretative framework, but any such framework should be in constant dialogue with the evidence. It would seem that the historical materialism that has recently dominated the scholarship on the Apion estate, with its teleological imperatives, lacks the necessary suppleness.

The present chapter takes a different approach. In the face of fragmentation and ambiguity—of the "strangely elusive quality" of the Egyptian peasantry[13]—it chooses description, for reading the traces of the texts and fashioning a structure of probabilities from these remains. It does not, for example, attempt to use the papyri to explain the legal codes (or consider the laws paradigmatic),[14] nor does it slot the Egyptian texts into one of Marx's constructs; what it seeks to represent is the complexity and diversity of the rural landscape of the late antique Oxyrhynchite, particularly the labor "panorama" of the vineyard. That this diversity existed is clear even if we examine only the media and rates of the payments that the estate received in the viticultural sector. In addition to the fixed cash payments for *apotakton chōriōn* and from other sources that appear throughout the dossier,[15] we find fixed-rate payments in kind in Text I, an account that also indicates share-

12. *P.Amh.* II 149, which concerns a *kēpouros* (gardener), is an exception, but it is ambiguous: what kind of labor relationship are we to see behind *hypo* in the phrase τοῦ ὑπ' ἐμὲ κηπίου (ll. 8–9; cf. p. 20 above)? *P.Oxy.* LXX 4789, which was published after Banaji's article appeared, would seem to provide the best support for his arguments. For the issue of *prochreia*, see pp. 84–85 below.

13. The words of Banaji (1999, 213).

14. "Explain the legal codes": as, e.g., Sirks 2008. Not paradigmatic: cf. the pithy assessment of Vera (1992–93, 320), who states, "To continue to write economic and social history starting from the codes is like crushing water in a mortar." Note also the emphasis on regionalism in, e.g., Marcone 1997. Generally speaking, it seems a mistake to *assume* the validity of a classical legal form for late antiquity; cf. the observations on the *coloni adscripticii* in Carrié 1997, 127.

15. Other sources: e.g., *P.Oxy.* LV 3804.135.

cropping.[16] Moreover, receipts from the dossier of Philoxenos *oinocheiristēs* ("wine steward") are suggestive of both types of payment in kind.[17] Indeed, a spectrum of arrangements should be expected in a society whose agents possessed varying resources and degrees of risk tolerance.[18]

Writing Labor

Community

It seems fitting to begin the descriptive process with perhaps the most striking feature of the exploitation of Apion vineyards: its communality, as attested by the accounts of the *pronoētai* and other texts. Communality in itself is not conspicuous in Egyptian agriculture;[19] what is remarkable is its application to viticulture. In the preceding chapter, I observed that *apotakton chōriōn* was invariably charged to collectives (principally *koina*, "communities"). It was there noted that only collectives (or wealthy individuals) could afford the high rates. But communality was not limited to *apotakton chōriōn*, which, as we have seen, may have governed imperial property only. It was also a feature of Apion vineyards,[20] including those that paid in kind (i.e., in must or wine). Thus, in *P.Oxy.* XVI 1896 (577), a *phrontistēs* and seven other men from the *epoikion* Leontos guarantee delivery of three thousand 8-*sextarius* measures of wine; this volume would seem to represent a fixed-rate payment for a total of 20 *arourai*.[21] The men refer to themselves as a *koinon* (l. 24), and the *phrontistēs* was their overseer.[22] It is tempting to draw a parallel with the earlier pittakiarch,[23] for example, but while the *phrontistēs conceivably* could

---

16. An abstract of Text I appears in appendix A.

17. *P.Oxy.* X 1326 and *SB* XXII 15320 are the published pieces of Philoxenos' dossier. Despite the odd number of vessels in 15320, I believe that both of these texts document fixed-rate payments, but share payments do appear in the unpublished part of the dossier.

18. Jördens (1999), e.g., elides local diversity.

19. For the earlier period, see Rowlandson 2005 on the *dēmosioi geōrgoi;* the Mamluke period, Kark and Grossman 2003 (especially at 226).

20. For cash payments to the estate, in all likelihood connected to its own vineyards, see, e.g., *P.Oxy.* XVI 1912.29, 85, LV 3804.135.

21. Using the rate of 150 *dipla* per *aroura* attested for Nekonthis in Text I. The estate standard was a *diploun* of 8 *sextarii*; cf. p. 190 below.

22. For the *phrontistēs*, cf. *P.Oxy.* LXII 4351 along with Mazza 2001, 129. He was clearly a middleman, probably an employee in some capacity, but no doubt well connected in his community. (A component of a *phrontistēs'* salary may be attested in *P.Oxy.* XVI 2024.7–8, 21–22; note also the *condicio* [legal status] of *enapographos* in *P.Amh.* II 149 and *SB* XVI 12484.) A thorough study of these individuals would be welcome.

23. For this figure, cf. Rowlandson 1996, 82.

have allotted vineyards,[24] Text I seems to indicate that the estate ultimately held individual teams of *geōrgoi* and *ampelourgoi* responsible for their parcels.[25] It also reveals, however, the (uncompensated?) transfer of wine from one group to another at the *epoikion* Nekontheos, which may denote collective action and solidarity.

What were these *koina*? This has been an issue of some debate, with Gagos and Van Minnen representing one extreme and Bonneau the other.[26] the former argue for endogenous communal organization and collective property, while the latter attributes the existence of these communities to external forces with fiscal and administrative objectives. Comparative evidence from the periods before and after ours would seem to support the fiscal argument: the *dēmosioi geōrgoi* ("public farmers") and cultivators under the *musha'* system both were working state land.[27] *Apotakton chōriōn* likely follows this pattern, but it is also clear that the *koina* could possess their own property,[28] and while it is possible to adduce fiscal and administrative interests in connection with nearly every document, we must ask if they are salient. When a *koinon* signed off on *P.Lond.* V 1764, an Oxyrhynchite advanced sale of wine from the last quarter of the sixth century, the economic interests of the parties, especially those of the seller-debtor "community," were likely at the forefront.[29] In short, Bonneau's research on this issue, though not without validity, fails to bear witness to the vitality of late antique rural communities.[30] In asserting that the *koinon* was "a grouping of persons brought together by their shared fiscal responsibility, a solidarity that was *more suffered than desired*" (1983, 513; my

24. If allotment occurred, it seems rather unlikely that parcels would have been distributed to different individuals each year, as under, e.g., Scottish *runrig* (for which see Whittington 1973, 542). I suspect that the Serenos who is expelling a *geōrgos* from his lease in *P.Oxy.* XVI 1941 is a *pronoētēs.* (This Serenos may reappear in *P.Oxy.* VI 914 of 486; if so, 914 provides a terminus post quem for 1941.)

25. Text I is abstracted in appendix A. Individual liability would seem to be confirmed by Benaissa, forthcoming, and by the receipts of the dossier of Philoxenos *oinocheiristēs.* In any case, I am reminded of the distinction between *affittanza a conduzione divisa* and *affittanza a conduzione collettiva* on modern Italian cooperatives.

26. Gagos and Van Minnen 1992 and Bonneau 1983.

27. *Mīrī* land (known as *kharajiya* in Egypt) in the latter case; cf. Kark and Grossman 2003, 223: "The ultimate title was vested in the state."

28. *P.Oxy.* VI 999.12 (617) [full ed. in Hickey, forthcoming]: π(αρὰ) τοῦ κοινοῦ τῶν γεωρῶν (ὑπὲρ) ἰδ(ίας) γῆς *vac.* σίτου κ(αγκέλλῳ) ρπδ[, "from the community of agriculuralists, for their own land, *cancellus artabai* of grain 184[" (a considerable amount given that it was *entirely* tax).

29. This text probably pertains to the Apiones, but other landowners (e.g., Flavia Anastasia) are also possible.

30. Cf. the assessment in Wickham 2005, 427–28 (though some of the Apion evidence is misinterpreted there).

emphasis), she has taken the "fiscal model" to an extreme (and perhaps an ironic one, given the role of her school in the rejection of earlier representations of Egyptian "serfdom").[31]

Tenure

Preliminaries. The question of tenure is a much more vexing one. In other words, what was the precise nature of the "payments" to the estate that appear in the our texts? The accounts of the *pronoētai* provide the most suitable point of departure.[32] These consist of two sections, the *lēmmata,* or "inflows," and the *analōmata,* or "outflows." The *lēmmata,* which were derived from the rent roll or *apaitēsimon,* consist of fixed (even fossilized) entries with sums of money (typically gold *solidi*) and/or amounts of commodities (usually *sitos* and never wine) from various parties; because they seem rarely (if ever) to have changed, they might be considered quotas or targets. Here is a sample entry: παρὰ Ψύρου ἀπὸ τοῦ αὐτοῦ, σίτου ἀρτάβαι δ, νομίσματος ∠ ιβʹ, "From Psyros from the same [*ktēma*], 4 *artabai* of grain, 1/2 + 1/12 of a *solidus*" (*P.Oxy.* LV 3804.51). The *analōmata* are the variable portion of the accounts (though they do have entries that are labeled "customary"). We encounter in them not only the expenses connected with running the steward's *prostasia* but also transfers of *lēmmata* to other administrative entities and concessions and other write-offs: for example, τοῖς γεωργοῖς τοῦ κτήματος Κοτυλεείου ὑπὲρ ἐνχόρτου ἀμπελῶνος ἀπὸ ἀρουρῶν ιγ ἄρουραι ιβ dʹ, τῇ ἀρούρᾳ α νομίσματος ῶ, νομίσματα η ςʹ, "[conceded] to the agriculturalists of the *ktēma* Kotyleeiou for vineland under grass, from 13 *arourai* 12 1/4 *arourai,* at the rate of 2/3 *solidus* per *aroura,* total 8 1/6 *solidi*" (*P.Oxy.* LV 3804.163). The objective of the *pronoētēs* was to balance the two sections of the account. As noted above, significant surpluses in grain seem rarely to have been an issue;[33] any excess gold was transferred to the estate banker (*trapezitēs*).[34] In the previous chapter, we saw that the accounts of the *pronoētai* present difficulties for determining the sources and natures (estate or nonestate property, private or fiscal/public) of *lēmmata.* The *lēmmata* are similarly

---

31. Keenan 2007 may be recommended for a balanced account of the rural settlements in Byzantine Egypt.

32. What follows is but a brief overview of these accounts; cf., further, Mazza 1998.

33. See p. 29.

34. For the recording of payments to the *trapezitēs,* see, e.g., *P.Oxy.* XVI 1911.212–17. As noted above (p. 50 n. 63), grain taxes were an expense of the *pronoētai,* while money taxes were paid on the estate level, out of the global *lēmmata.*

uninformative regarding tenure,[35] but when we read an entry like "From the community of the agriculturalists and vineyard specialists, for the ground of the vineyard specialists, *solidi* 13½,"[36] it is difficult to view this cash "charge" (and most of the others in the accounts) as anything other than a money rent. It is rather unlikely that such entries conceal income in kind, that is, that they represent valuations of the commodities produced on the land. This is actually a matter of broad significance, since bookkeeping on the Apion estate has generally been considered to have been rather rudimentary;[37] commodity valuation would mean that estate accounts were rather more sophisticated than has been thought. It would indicate, for example, that payments by *pronoētai* to estate bankers had a fictive element,[38] and as a result, some views concerning the nature of the Apiones' income would need to be reconsidered.[39] It would also bring us a step closer to demonstrating that accounts like Text I, the pronoetic *logoi,* and the *oinocheiristēs*' account all interlock. The extant evidence will not, however, stretch this far. Two work contracts for estate *pronoētai* are preserved.[40] Though each *pronoētēs*' "duties are described in detail,"[41] no responsibilities connected with wine or other commodities (besides *sitos*) are even intimated, not even oversight.[42] Instead it is clear that a *pronoētēs*' duties were limited to the collection of grain and coin and that he was to forward the net amount of his collections to the appropriate officials.[43] Indeed, the potential benefit from or purpose for fictively in-

35. In them, *phoros* ("rent") is rarely used of land (as opposed to date palms, mills, dovecotes, bakeries, oil presses, landings, and sheep). The exceptions are *P.Oxy.* XVI 1911.87 (*edaphos*), 122–23 (unclear), 1912.137–39 (*apotakton chōriōn*), XVIII 2195.174 (*mega gēdion*), XIX 2243a.3 (*lachania*), 47 (*gēdion*), 53 (*pōmarion*), LV 3804.159 (*gēdion*); the impression is that it was not used of "generic" vineland and arable land. *Ekphorion* (a rent in kind, usually) does not appear in these accounts.

36. *P.Oxy.* LV 3804.135: παρὰ τοῦ κοινοῦ τῶν γεωργῶν καὶ ἀμπελουργῶν ὑπὲρ ἐδάφους τῶν ἀμπελουργῶν νομίσματα ιγ ∠.

37. Cf. Rathbone 1991, 402 ("apparently no more than unintegrated records of rents or taxes received or due and of expenditure made") and Mickwitz 1937, 583. Some rehabilitation was undertaken by Mazza (1998), and my own experience with the corpus suggests that additional exploration would prove fruitful.

38. Rea's explanation of *ektakta trygēs* (*P.Oxy.* LV 3804, p. 125, 96n.) has the same implications, but he does not pursue them.

39. See Hickey 2008, 90 (with refs.), emphasizing the family's "predilection for an income in gold."

40. *P.Oxy.* I 136 (= *W.Chr.* 383) and LVIII 3952. Note also Benaissa 2007.

41. *P.Heid.* V, p. 137, in description of *P.Oxy.* I 136.

42. As we have seen above, some supervision of wine production seems to have been performed by estate *chartularii* (cf. also *P.Oxy.* XVI 2059).

43. Cf., e.g., *P.Oxy.* I 136.19–22. In these lines, the *pronoētēs* contracts himself "for the collection and payment of everything to your excellence or its representatives, namely, the *sitos* to the public boatman of its glorious house and the cash to the most illustrious banker of the same glorious

cluding commodities in the records of his payments is elusive.[44] Furthermore, none of the extant rent rolls, the documents on which *pronoētai* based their collections,[45] include rents in forms other than grain and money. Regarding wine, Text I and receipts from the dossier of Philoxenos reveal that *ampelourgoi* and *geōrgoi* paid their wine directly to the wine steward (*oinocheiristēs*), and this estate official seems not to have reported the yield to the relevant *pronoētai*—at least, there is no evidence for this.

To propose that the cash *lēmmata*, or the majority of them, are best understood as rents is to suggest, of course, that the corresponding lands were exploited under some form of tenancy. The presence of entries in the *lēmmata* with sums for *ektakta trygēs* would seem to support this; supplementary payments to the landlord—be they termed *synētheiai* ("perquisites"), *heortika* ("festal gifts"), *ektakta* ("special" dues), or the like—were a common feature of tenancy agreements.[46] *Trygē* is furthermore indicative of the vintage, and it is noteworthy that four of the six unique attestations of *ektakta trygēs* occur at sites with *apotakton chōriōn*;[47] *ektakta trygēs* is also assessed against collectives in at least four of the references.[48] But there are apotactic sites (e.g., the *epoikion* Erotos; see table 2.1.10) where these *ektakta* do not appear,[49] and, moreover, they need not even indicate tenancy; they could feature in labor contracts, specifically in leases of work.[50]

---

house" (εἰς τὸ πάντα εἰσπρᾶξαι καὶ καταβαλεῖν ἐπὶ τὴν ὑμῶν ὑπερφύειαν ἤτοι ἐπὶ τοὺς αὐτῇ προσήκοντας, τοῦτ᾽ ἔστιν, τὸν μὲν σῖτον ἐπὶ τὸν δημόσιον ναύτην τοῦ ἐνδόξου αὐτῆς οἴκου, τὸ δὲ χρυσικὸν ἐπὶ τὸν λαμπρότατον τραπεζίτην τοῦ αὐτοῦ ἐνδόξου οἴκου).

44. An attempt to use the *pronoētēs*' account to determine the profitability of the *prostasia* seems improbable, given that the waters were already muddied with fiscal charges.

45. Cf. *P.Oxy.* I 136.17–18.

46. In general, see Herrmann 1958, 118–21. In the late antique Oxyrhynchite, *ektakta* are specifically referenced in, e.g., *PSI* VIII 920.6. In lease contracts, the payments are typically in kind (dates, figs, etc.), whereas in the pronoetic accounts, they are in money, often supplemented by grain. Table 3.1 provides an overview of the Apion evidence.

47. If table 3.1.6 concerns Psempekle, the number increases to five.

48. The individuals apparently charged in *PSI* VIII 954.12 (table 3.1.5) were probably "substantial peasants." The liable parties in 3.1.1 are uncertain; I presume that reference is being made to the preceding entry (*P.Oxy.* LV 3804.103: 158.4 *art.* and 7.5 *sol.* [according to the image; ed. pr. is incorrect] due from land connected to two individuals, to be paid through a collective).

49. This raises the question of scribal consistency, but at the *epoikion* Erotos at least, it does not appear that *ektakta trygēs* are classified as something else. It is tempting to interpret *ektakta trygēs* as an umbrella term for various vineyard-related charges. With one exception (*PSI* VIII 954.5–7), it does not appear alongside the *phoros phoinikōn* ("rent of date palms"; for vineyards, dates, and *ektakta*, cf. *P.Oxy.* XIV 1631.20–25 [280]) or the *phoros peristereōnos* ("rent of a dovecote"; vineyards, being artificially irrigated, would have been deprived of the natural fertilization provided by the inundation). The payments of grain in the *ektakta* (table 3.1.2, 4–6) could indicate arable land attached to vineyards; cf., e.g., *P.Hamb.* I 68 (Aphrodite, 548–49 or 563–64).

50. Cf., e.g., *P.Oxy.* XIV 1631. These contracts are discussed further below (pp. 79–81).

TABLE 3.1. *Ektakta Trygēs*

Reference; Site	*Ektakta Trygēs*
1. *P.Oxy.* XVI 1911.33 (*SB* XXIV 16324.34)/ LV 3804.104; (individuals at? collective at?) Tarouseb(t)	5.083 *sol.*
2. *P.Oxy.* XVIII 2195.17; Tillonos[a]	3 (*artabai*) *sitos* + > .047 *sol.* + 2,700 myriads of *denarii*
3. *P.Oxy.* LV 3804.96/*SB* XXIV 16324.25; Loukiou	.417 *sol.* + 400 myr. *den.*
4. *PSI* VIII 954.7; Skytalitidos	6 *cancellus art. sitos* + 4.875 *sol.* + 7,800 myr. *den.*
5. *PSI* VIII 954.12; individuals?	2 *canc. art. sitos* + 2+? *sol.*[b] + 2,609 myr. *den.*
6. *PSI* VIII 954.40; Mikras Paroriou? Psempekle?	> 2 *canc. art. sitos* + 6.375 *sol.*[b]

[a]See appendix B, s. ΤΙΛΛΩΝΟΣ.
[b]Read on original.

Resource sharing. Instructive for the issue of tenure is the examination of resource sharing between the estate and its labor force, given that under most forms of tenancy, the lessee contributes more than his labor.[51] Take, for example, *P.Cair.Masp.* I 67104 (530), a lease from the village of Aphrodite in which the *magnificentissima* daughter of a *gloriosissimus et magnificentissimus comes* lets a vineyard to a certain Aurelius Ioannes.[52] In this contract, which had a term of ten years, the tenant Ioannes had the following duties (beyond paying his annual rent of 120 7-*sext. angeia/ar.*):

- Winter and summer irrigation.
- All vineyard work (*ampelourgikē ergasia*), any expenses of which were to be borne by Ioannes. Likewise, Ioannes was required to use his own animals.
- Furnishing vessels for his rent.
- Ioannes also agreed to farm some adjacent pieces of land, as well as some dry vineland (*chersampelos*), according to terms accepted by a former tenant. In addition, he assented to paying the date palm rent (*phoros tōn phoinikōn*) and other extras (*ta alla ethima,* literally, "the other usual things") per the former tenant's lease. Finally, Ioannes

51. For resource sharing under tenancy in antiquity, cf. Foxhall 1990 (which does not, however, mention the lessee's human capital, i.e., his "know-how"; cf., e.g., Hallagan 1978, 348–53). "Under most forms": the share tenants of the Fayum come to mind as an exception; see Jördens 1999, 124.

52. For the lessor's identity, cf. Ruffini 2008, 190.

agreed to enclose the dry vineland over two indictions, with the landlord supplying the wood and other expenses.[53] During these improvements, the dry vineland would be rent free.

This text not only intimates the inherent variability of leases (the terms of which were dependent on the parties involved and their resources, the character of the land, and local custom, among other factors); it also sheds light on estate practice. If we convert 67104's rent in kind into one in cash, the result is 2.6 *solidi* (on the public standard) per *aroura*.[54] This is close enough to the 3 (public) *solidi* per *aroura* that were paid under *apotakton chōriōn* to suggest similar responsibilities and resources for the collectives working these properties; at any rate, one would not expect employees to make such contributions.[55] Text I, in turn, reveals that the teams of *geōrgoi* and *ampelourgoi* who paid the estate *in kind* were responsible for transport and cellaring charges,[56]

53. Schnebel (1925, 256) argued that the wood was for vine props, not fencing. For the use of a palisade (*phragmos*) around (new) vineyards, see Kloppenborg 2006, 158–61.

54. For this calculation, I have used a datum from the Apion estate account *PSI* VIII 953, in which a commutation rate of 30 *dipla* per *solidus* is attested (1.57). A *diploun* of 8 *sextarii* was standard in this account (cf. p. 124 below), and I assume that the private (estate) gold standard was intended. For 67104, I have inferred the goldsmith's standard (cf., e.g., *P.Michael.* 45.39–41), and I interpret the gold standards as does Banaji 1998.

55. That Ioannes' resources and responsibilities invite comparison with those of a collective is more remarkable. He is known from five other texts: *BGU* XVII 2711 (dated VI); *P.Cair.Masp.* II 67143 (dated VI), III 67327 (540); *P.Mich.* XIII 671 (mid-VI); and *SB* XX 15018 (554). The *Sammelbuch* papyrus reveals that he was one of the *prōtokōmētai* of Aphrodite, along with Aurelius Phoibammon, son of Triadelphos (with whom he is also involved in *BGU* 2711 and *P.Mich.* 671). *P.Cair.Masp.* 67327 preserves his hand; he could write fairly well.

56. Those making share payments in Text I paid a fixed cellar charge of 4 *dipla* per team and the *naulon paktonariou*, "skiff/tender freight charge" (for these vessels, cf. *P.Oxy.* XLVIII 3406), which ranged from .67 to 1 percent of the total volume of wine delivered (.92 percent is the mean; $N = 7$). The cellar charge was presumably assessed to cover the cost of storing the teams' share of the must/wine, but it cannot be excluded that it was a payment for the storage of the estate's share until it was deemed "good"; see further below, p. 96 n. 20. No charges for terrestrial transport appear at Text I's "share" *epoikion* (the name of which is not preserved). The use of skiffs for transport suggests that it was close to canals or at least to waters only navigable by small boats when the wine was to be transported (probably October through December; cf. the delivery dates in the forthcoming dossier of Philoxenos).

The teams at Phankona and Nekonthis, which paid a fixed quantity of wine (or rather, must; cf. p. 77 n. 90 below) per *aroura*, were assessed for *kamēlarioi* ("camel drivers"; cf. *CPR* XIX 66 and refs.) and the *naulon ploiou*, "ship freight charge." At Phankona, the former assessment is consistently .19 percent of the total volume of must due for the indiction and .2 percent of the actual volume of must delivered. At Nekonthis, the percentages range from .13 to .16 percent of the sum due for the indiction and from .19 to .23 percent of the actual volume delivered; one team pays nothing at all, perhaps because they had a bad year. These small payments probably represent gratuities for the estate's *kamēlarioi* (whose salaries are attested in *P.Oxy.* XVI 1911.156 and LV 3804.238); see Hickey 2001, 119–20. The *naulon ploiou* at Phankona is consistently 1.94 percent of the amount of must due while traction would seem to be a contribution that most anyone *could* have made.[57] Reeds may also have been an input from labor: on no occasion is the estate attested purchasing reeds for the maintenance of productive vineyards (something that one would expect even under direct exploitation);[58] estate purchases are documented only for *neophyta chōria* (cf. table 3.2).[59]

Vessels and irrigation inputs are matters that are less clear.[60] It would seem rather unlikely that those paying cash rents were supplied with vessels for their wine yield, and, indeed, sites attested with such leases coincide with those receiving vessels from the estate only twice.[61] Regarding those furnishing wine or must, *P.Oxy.* XVI 1896, which is probably a fixed-rate agreement for 20 *arourai*,[62] makes no mention of the supply of vessels;[63] an ineditum, however,

---

and 1.97 percent of the quantity delivered; at Nekonthis these percentages range from .89 to 1.6 percent and from 1.86 to 2.06 percent, respectively. These rates suggest (unsurprisingly) that the must was not transported out of the region; cf. Hickey 2001, 122–23. The use of *ploia* (as opposed to *paktonaria*) may indicate that the watercourse nearest to Phankona and Nekonthis was navigable by large vessels at the time when the must was to be moved (probably soon after the vintage); this watercourse was not adjacent to the vineyards, however, and beasts of burden were required to bring the must to the boats.

57. In certain sureties for *enapographoi*—who, as we shall see, were typically wage laborers—the property of these individuals is (stereotypically) identified as *ktēnē* ("domestic animals") and *pasa hē aposkeuē* ("all the household goods"); cf. *P.Heid.* III 248.7; *P.Mert.* II 98.4–6; *P.Oxy.* I 135.17–18, XLIV 3204.14–15, LXX 4790.24–25, 4791.17–18; *SB* XVIII 14006.21–22. Note also *P.Oxy.* XVI 1913.1–5 and XVIII 2195.139–40, though the first of these perhaps implies that estate animals were typically used. In the model *misthapochē SB* XX 15027 (a combined receipt and tenancy contract; reed.: *APF* 45 [1999]: 39–46), moreover, the lessor furnishes use of his animals. According to Jördens (1999, 124, 138), supplying animals was beyond the means of Arsinoite share tenants. In nineteenth- and twentieth-century Egypt, there was a spectrum of sharecropping arrangements; *mukhammas* croppers (who received a one-fifth share) did not furnish their own animals (or anything else besides their own bodies). The expense and care of animals on the *ʿizab* were shared between the landowner and *ʿizbah* worker; see Richards 1978, 516–17. Some late antique contracts distinguish animals from their maintenance; cf. *SB* XIV 12050.21 and XVIII 14001.19.

58. For reeds in viticulture in general, see Schnebel 1925, 255–61, with the additions of Rathbone 1991, 187–88, 248–49. "One would expect": reed beds were attached to many vineyards, but self-sufficiency seems unlikely given the huge quantities needed (cf. *P.Bingen,* p. 446). The Appianus estate, which exploited its vineyards at Theadelphia directly, was not self-sufficient in reeds (Rathbone 1991, 249; Kehoe 1992, 108); it acquired reeds from tenants who had leased beds from the estate (Rathbone 1991, 187–88) and received them from outside the *phrontis* (either from central stores or through purchases; so *P.Bingen,* pp. 443, 445).

59. "It was usual in the case of new vineyards for the lessor to supply reeds and binding material," notes Kloppenborg (2006, 327 n. 179).

60. For the estate's economic relationships with potters, see Morelli 1997, 16–22, 28–29, and Hickey 2001, 100–19. For advanced sales of pottery, see *P.Eirene* II 28 and 29 (with refs.).

61. *P.Oxy.* XVI 1912.29, 127–28 (Papsau), XVI 1911.181–92 (+ *BL* XI 154–55), LV 3804.218–19, 3805.87–88 (Tarousebt).

62. See p. 65 above.

63. The *geouchika sēkōmata* mentioned (in l. 19, e.g.) are decanting measures (so Mayerson 1998b). Incidentally, in 1896.25, read τ̣ῆς ῥύσε[ω]ς ὡς πρόκ(ειται) for]η̣ ̣ [̣ ̣ ̣ ̣ ̣]ς̣ ὡς πρό[κ(ειται)].

TABLE 3.2. Purchases of Reeds

Text	Bought for	Area	Cost	Notes
1. *P.Oxy.* XVI 1911.175–77	—*ktēma* Kotyleeiou —*ktēma* Tarouseb(t)	100 bundles, divided respectively: —1.25 *ar.* —.25 *ar.*	Total: 2 *sol.* less 9 *ker.* = 2 *sol.* on the private standard w/*rhopē* = 1 *sol.*, private standard w/*rhopē*, for 50 bundles; if distribution proportional to area, then .75 *ar./sol.*, private standard w/*rhopē*	—reeds are new —for *neophyta chōria* —purchased on Mecheir 16
2. *P.Oxy.* XVI 1912.152–54	—*epoikion* Papsau —ἐν τῇ Θρυείτιδος καὶ ἐν ἐποικίῳ Κλέωνος ἐν τῇ μηχανῇ Ψανσηννέ —*epoikion* Graeidos —*epoikion* Chenetorios	respectively: —2 *ar.* —1 *ar.* —1 *ar.* —1.5 *ar.*	explicitly 2 *sol./ ar.*; total, read questionably, of 11 *sol.* less 39.5 *ker.*; undoubtedly 49.5 *ker.* appeared, making *sol.* on private standard w/*rhopē*	—reeds are new —for *neophyta chōria*

intimates that amphorae were provided by the estate under this kind of agreement,[64] and this may be confirmed by further inedita in the dossier of Philoxenos *oinocheiristēs*.[65] The Philoxenos texts additionally suggest the possibility that those making share payments received vessels as well;[66] the large amphorae production numbers attested in the dossier may also support this.[67]

64. See Benaissa, forthcoming.

65. If amphorae produced at a site were indeed (mostly) expended at that site.

I also note that fixed-rate sites seem to have delivered immediately after the vintage (cf. *P.Oxy* XVI 1896; Benaissa, forthcoming). In sales in advance of wine with vintage delivery, the buyer usually supplies the vessels (cf., e.g., Jakab 1999, 39–40).

66. There is some overlap between the share sites and the amphorae production sites. I propose below (p. 77 n. 90) that share payments were delivered not at the vintage but some months later. In advanced sales of wine with delivery after the vintage, vessels were sometimes provided by the buyer (cf., e.g., Jakab 1999, 40–41). Note that advanced purchases of *vessels* invariably specify delivery at or just before the vintage (cf. *P.Eirene* II, p. 161).

67. Some caution is in order, however, when assessing vessel production figures. For one, an order of amphorae was not expended completely at the press (on the contrary, cf. Rathbone 1991, 253, and Empereur 1993, 42). Moreover, vessels might be used for purchased wine (e.g., that acquired through advanced sales). Another possibility under share tenancy is that the parties each provided the vessels for their own shares. For further discussion of this matter, see Jördens 1986b, 110.

As for irrigation, the estate retained responsibility for capital improvements and, typically, repairs to water systems,[68] which, as I indicated above, were inherently fiscal.[69] Some contemporary Oxyrhynchite leases, however, place the responsibility for embankment work on the tenant.[70] In the Roman period, the work itself of supplying water to crops seems to have been the tenant's responsibility as a rule,[71] and this trend appears to continue in late antique Egypt,[72] including the sparsely documented Oxyrhynchite.[73] Yield calculations based on the sharecropping portion of Text I suggest the possibility, however, that the *oikos* paid for irrigation under the arrangements current at that site (and perhaps at sharecropped vineyards throughout the estate). As often is the case when we work with ancient data sets, some assumptions are required. For one, contemporary data for wine yields are all but nonexistent; there is only *P.Cair.Masp.* I 67104 (discussed above), which furnishes the following data point: 33.3 hectoliters per hectare.[74] The areas of the share vineyards in Text I

68. Some examples of estate involvement/expenditure: *P.Oxy.* I 134 (blocks of stone for a cistern), XVI 1911.160–65, 169–74 (axles for *mēchanai*), 166–68 (stone blocks for a cistern), 1913.19–23 (wages for artisans working on various *mēchanai;* half of the wage of the *chalkeus* [smith], however, was paid by the *geōrgoi* of the site), XVIII 2197 (numerous allocations of bricks for, e.g., cisterns), XIX 2244 (extensive axle distributions), LV 3804.213–17 (expenses connected with the construction of a new cistern), 221–22 (wages for a *potamitēs* or navvy working on a cistern), 223–24 (wages for *potamitai* reembanking a damaged vineyard), 249 (discharge trough for a *mēchanē*), LXVII 4537, 4538 (measurements for cistern construction), LXVIII 4704 (wages for *potamitai* working in *chōria*). For the cost of a *mēchanē,* cf. Hickey 2007, 294.

69. P. 50. Cf. also, e.g., *P.Oxy.* VII 1053.12–20 (payments from various independent villages to the estate, presumably to fund [or reimburse] dike maintenance. Incidentally, in 1053.2, καὶ κοιν(ωνῶν) should probably be read for καὶ καιν(οποιοῦ)).

70. Cf. *P.Berl.Zill.* 7.25–26 (574): ποιούμ(ενος) [my suggestion; ed. pr.: ποιοῦμ(αι)] καὶ τὴν ἀναβολὴν τῶν ναυείων, "I will also perform the raising of the irrigation earthworks." For similar statements, see *P.Mich.* XI 611.18–21 (412); *P.Michael.* 34.8 (VI); *P.Oxy.* LXIII 4390.23 (469; Apion lease); *PSI* VIII 881.4 (VI); *SB* XXIV 16128.2–4 (V), XXVI 16722.8 (late VI; Apion lease?). The reading χ̣ω̣μ̣α̣τ̣ι̣σ̣[μούς in *BGU* XVII 2682.28 (481, Hermopolite) seems all the more suspect in light of the contract's terms, but cf. *SB* XVIII 14001 (486, Arsinoite).

71. Cf. *P.Col.* X, p. 100, where it is noted that the "lessor bearing responsibility for irrigation is rare." The terminology of irrigation—*antlēsis, ardeia, potismos, hydroparochia,* etc.—is addressed by Bonneau (1993, 209–20) but would, I believe, repay further study.

72. *SB* XX 15027 (see n. 57 above) and *T.Varie* 14 (Herakleopolite) come to mind as exceptions. In those sharecropping agreements in which the tenant receives a small share (e.g., 25 percent) and in which no mention is made of his involvement with irrigation (cf., e.g., *BGU* XVII 2682 from the Hermopolite), the landlord was presumably the responsible party. See further discussion below.

73. Cf. *P.Oxy.* LV 3803.13 (411).

74. Not 26 hectoliters per hectare, as has Rathbone (1991, 247 n. 51), who is apparently unaware that wine was sharecropped at a ratio of 1:1 in the Byzantine Antaiopolite. Using this figure (in part), Rathbone estimates ca. 25 hectoliters per hectare for the yield at Appianus' Theadelphian *phrontis.* Cockle (1981, 96) states that modern Egypt *ought* to be producing ca. 28 hectoliters per hectare. In one contemporary viticultural manual, however, "serious overcropping" is defined as "> 100–150 hl/ha" (Jackson 2008, 93), while the "base yields" (= unfermented grape juice) for lower grades of

are also lacking, but there are some figures from the fixed-rate sites. If we apply their mean (2.25 *ar.*; but $N = 6$, $SD = 1.41$) to the seven vineyards at the share site—that is, if we posit a total area of 15.75 *arourai* (4.34 ha) for the share site[75]—and if we assume the standard estate wine amphora of 8 *sextarii*,[76] we only end up within 20 percent of the figure from 67104 (viz., 26.6–40 hl/ha = the shaded area on fig. 4) with a 2:1 or 3:1 division; 3:1 calculates to a remarkably close 33.4 hectoliters per hectare.[77] An even division of the yield becomes possible only when we maintain that the share site's vineyards totaled 19.74–29.54 *arourai* (5.44–8.14 ha). Given the uncertainty of our yield and area assumptions, we cannot, of course, exclude this possibility, but regional trends suggest that we perhaps should. Jördens has observed a "social decline" from south to north in sharecropping leases, in which the tenant's lot generally deteriorates the farther north one goes.[78] To the north of the Oxyrhynchite in the Arsinoite, wine was typically sharecropped at 3:1,[79] and the landlord—given his large quota and the fact that contracts generally do not speak of irrigation (yet do bother to note hand tools like the *xois* furnished to the tenant)[80]—was responsible for supplying water to his vineyard. To the south, in the Hermopolite,[81] share tenants typically received half of the yield, but this was to be split with another individual. In most cases, the contracting tenant himself was an *ampelourgos*—this is either explicit or clear from the description of his work[82]—while an irrigator (*hydroparochos*) or *geōrgos* was his assistant, as in *BGU* XII 2175.4–5 ("and that I [the

French wine start at 80 hectoliters per hectare (Coleman 2008, 45); i.e., Egyptian vines presumably have much greater productive *potential* than Cockle indicates. For the comparison of ancient and modern data for wine yields (specifically with regard to ancient figures that seem unrealistically high), cf. Bergquist 1992, 119–22.

75. As an alternative approach, one could assume that the ratio of the landlord's yield to vineyard area at Text I's Nekonthis is identical to that of the *geouchikon meros* to vineyard area at the share site. This results in a vineyard area of 16.57 *arourai* (4.57 ha) at the share site ($M = 2.37$ *ar.*). The area figures for the estate vineyards in *P.Oxy.* XVI 2051r are smaller than the Text I mean(s): 1 *aroura* (× 2), 1.5 *arourai* (× 2).

76. Cf. p. 190 below.

77. A 2:1 division stays within 20 percent of 67104 with a total vineyard area of 14.77–22.12 *arourai* (4.07–6.1 ha); 3:1, 13.16–19.67 *arourai* (3.63–5.42 ha).

78. *P.Heid.* V, pp. 256–58.

79. *P.Ross.Georg.* III 51 (631) is an exception (2:1).

80. But note the mentions of canal work in *SB* XVIII 14000 and 14001. In *P.Vind.Sijp.* 10, the landowner lends the cropper money to rent animals for the irrigation machinery; the cropper is compensated for their fodder by the produce of a plot near the vineyard. For the *xois*, or pruning knife, cf., e.g., *SB* XVIII 13997.6–7.

81. The Herakleopolite evidence is nonresponsive for the issue under consideration.

82. Often the *ergasia* (work) is called *ampelourgikē* (viticultural) outright; when it is not, references to *kalamostasia* (setting vine supports) and *kalamourgia* (reed work), e.g., are more than suggestive.

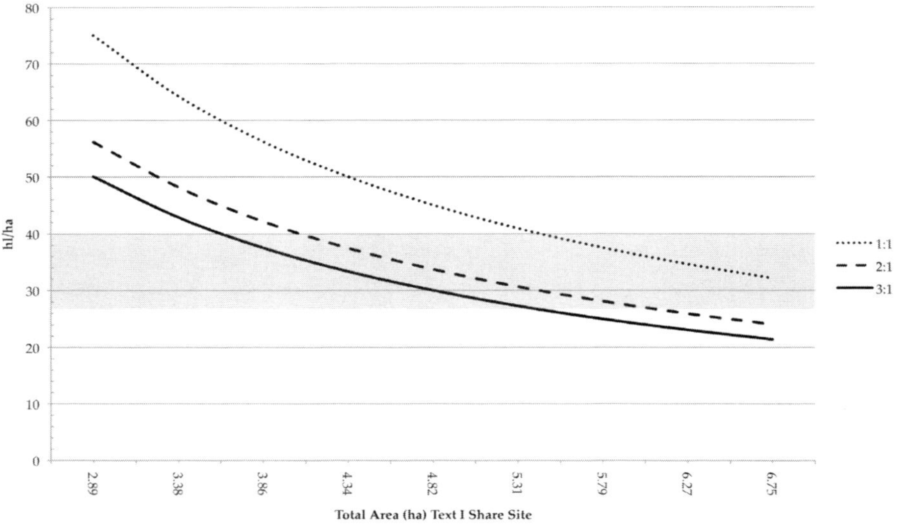

Fig. 4. Cropping quota-yield scenarios (Text I share site). (Graphic by the author.)

ampelourgos] along with the irrigator receive the other half share [of the yield] in exchange for our toil");[83] "generic" *geōrgoi* were no doubt chiefly responsible for irrigation.[84] That the watering of the vineyard and the *ampelourgos*' work within it were each valued at 25 percent of the yield is clear from those contracts that depart from the norm: in agreements in which no mention is made of irrigation, the *ampelourgos* receives a quarter of the wine,[85] while a cropper who takes responsibility for both types of work receives 50 percent.[86]

The account entries for the share site in Text I are rather suggestive of the arrangements attested in the Hermopolite contracts:[87] every one of them begins πρὸς ["due from"] *personal name* (in the accusative) ἀμπελουργὸν *personal name* (in the genitive) γεωργοῦ; the latter figure was undoubtedly

83. καὶ ἐμὲ μετὰ τοῦ ὑδροπαρόχου δέξασθαι τὸ ἄλλο ἥμισυ μέρος ἀνθ' ὧν ποιούμεθα καμάτων. (The ed. mistranslates ὑδροπαρόχου as *Kanalarbeiter.*) To my knowledge, an *ampelourgos* appears as the partner only in *P.Stras.* V 486.10–11 and *SB* XIV 12050.24–25; cf. also the deletion in *P.Vind.Sal.* 9.9–10. In all of these cases, the work of the principal party was irrigation.

84. Cf. the duties of the ⲟⲩⲟⲉⲓⲉ (*geōrgos*) in the Hermopolite share lease *P.Ryl.Copt.* 159 (VI–VII).

85. Cf. *BGU* XVII 2682 (481); *P.Coll.Youtie* II 89 (485); *SB* XXII 15493 (435).

86. Cf. *SB* XX 14416 (VI, Antinoopolite). *P.Oxy.* XII 1590 (with Rowlandson 1996, 230) suggests that irrigators could take on the work of the *ampelourgos.*

87. The employment of the "lease of work" (on which see p. 79 below) provides another connection between the labor systems of the Oxyrhynchite and Hermopolite. Cf. also the assessment in Fikhman 1973, 17–18, with its "broadly similar conclusions" (though also noting important differences).

responsible for irrigation.[88] This need not indicate, however, a formal partnership between the *ampelourgos* and *geōrgos;* in the Hermopolite examples, the irrigator/*geōrgos,* though a beneficiary of the contract, is not one of the parties to it. Indeed, Jördens believes that the Hermopolite *geōrgoi* were probably employees of the landlord.[89] A similar arrangement could have applied at the Text I share site as well, but only if we wish to argue for a 1:1 division of wine, which was not indicated by the model presented above. It is also possible that the irrigator was not paid with a share of the yield, under which circumstance a 3:1 division (i.e., 25 percent for the *ampelourgos*) presumably applied (see further below). Regarding this question, it may well be relevant that those working in Text I's *fixed-rate* vineyards received a share of must in the vicinity of 25 percent.[90]

In light of their contributions of resources—even if they are less than we might expect or less than the norm (whatever that means)—it seems unavoidable that the *ampelourgoi* making share or fixed-rate payments in kind to the estate were tenants; it may be noted that the contracts governing their "much too poor" and downtrodden counterparts in the Fayum were certainly leases, if only de jure.[91] As Jördens' hypothesis about the Hermopolite irrigators suggests, however, there is a fine line between cropping and wage labor; in fact, the croppers in *CPR* XVIIA 6 refer to their quarter share of the yield as a wage (*tous misthous,* l. 16).[92] This ambiguity, combined with other (de

88. At Nekonthis and Phankona, the fixed-rate sites in Text I, the team structure is less consistent, but *geōrgoi* are part of the production groups in all but one case.

89. *P.Heid.* V, p. 256: "Here [in the Hermopolite] it is stipulated to the tenant that he has to provide for another person—*probably an employee of the landlord* [my emphasis]—from his share."

90. At Nekonthis, the amount due from each group is the equivalent of 23.78 hectoliters per hectare, or 71.2 percent of the yield of a vineyard producing 33.4 hectoliters per hectare. It should be remembered, however, that the fixed-rate teams were subject to terms that varied from those of the croppers. The differences in their extra charges have already been noted (n. 56), and the evidence of which I am aware (published and unpublished) leads me to hypothesize that those paying a fixed rate delivered must (cf. n. 65), while the croppers had to furnish wine (i.e., fermented must). See also p. 96 n. 20 below. The landlord's share at cropping sites was no doubt calculated on the must yield, and spoilage before delivery may account for some of the arrears that Text I attests at the share site. Cf. the *anchara* system in Andhra (India), under which shares were determined from an *estimate* of production (Satyanarayana 1985, 112–13).

91. Cf., e.g., *P.Vind.Sijp.* 10. "Much too poor" is the characterization of Jördens (1999, 124), who also notes "the low social standing of Arsinoite share croppers" and that they stood "in an especially high dependence on the landlord."

92. A point emphasized in Banaji 1997, e.g.; note also the expression "in exchange for the work that we perform" (ἀνθ' ὧν ποιούμεθα καμάτων and the like), which references the yield in Byzantine share contracts. Such locutions presumably would not have passed muster in an Islamic agreement; under Muslim jurisprudence, a laborer cannot be paid with the fruits of his labor. See Firestone 1975, 5–8 (with refs.) for further discussion of this issue, which illuminates the ambiguous nature of sharecropping.

facto) circumstances of Apion tenants—the decoupling of law and practice "on the ground"[93]—has contributed to the creation of divergent narratives about labor deployment on the estate. Rents furnish a good example of this decoupling. It appears that their form could be converted, certainly from cash into kind and perhaps vice versa.[94] Even, however, under *relocatio tacita* (the tacit renewal of a lease), which likely prevailed on the estate,[95] the terms of an expired but continuing lease could not have been altered; a new lease should have been required.[96] The very existence of a fixed rent roll implies that this did not happen.[97] Laws could be disregarded, of course,[98] and tenants might even prefer this: a conversion to a rent in kind would have relieved them of some of the burden of marketing and, ceteris paribus, would probably have been desirable. But what if the new terms seemed unfavorable or prices were extraordinarily volatile?[99] Or what if the *geouchos* later wished to return to a

93. Cf. n. 14 above.

94. A transformation from cash into kind would have been a simple matter; the *pronoētēs* merely needed to discharge the cash sum due in his *analōmata.* A "write-off" of this sort occurs in *PSI* VIII 954.56–57, where I suggest reading (ὑπὲρ) οἴ(νου) κν(ι)δ(ίων) φιϛ δοθ(έντων) [τῷ ο]ἰνο-χειρ(ιστῇ) δ(ιὰ) τ(ῶν) ἀπὸ Φαγκόνα (καὶ) προσεν̣[εχθέντων εἰς τὰ ἀναλώματα τούτου τοῦ λόγου] κατὰ τὸ ἔθ[ος (καὶ) ἐπὶ τῆ]ς θ ἰνδ(ικτίονος) [νο(μ.) *x*, on the basis of a photograph kindly furnished by Rosario Pintaudi. (The length of the lacuna following προσεν̣[εχθέντων is unknown; abbreviation presumably occurred.) Phankona happens to be known from Text I, where it is one of the two sites that delivered a fixed quantity of wine per *aroura,* specifically 172 *dipla* per *aroura,* if I have read the number of *arourai* correctly. (The numeral 3 is clear, but fractions may have followed.) The total number of vessels due from the settlement is identical to the sum mentioned in *PSI* 954; this is surely not a coincidence. (In Text I, Phankona is said to owe a total of 515 *dipla,* but the figures in the subentries suggest that there has been an addition error and that 516 *dipla* were intended. For attestations of the equivalence of the *diploun* and *knidion* [the measure in *PSI* 954], see appendix C, s.vv.) For another (probable) rent conversion, see p. 99 n. 35 below.

95. Cf. p. 20 above.

96. See Kehoe 2007, 125–26. I am not aware of any late antique lease in which the form of the *principal* rent was left to the choice of the tenant or landlord, but note *P.Coll.Youtie* II 89.20–21 (called "unique in Byzantine contracts" by its ed.; apparently this is still the case). Roger Bagnall reminds me, however, of the Kellis agricultural account book, in which the *pronoētēs* was ready "to accept any commodity or cash in place of any other commodity or cash" (*P.Kellis* IV, p. 62). For comparable practice in the British Isles, see Mossner and Ross 1987, 151–52 (letter of Adam Smith to Lord Hailes).

97. The lease contract of indefinite duration also comes to mind; cf. Jördens 1999, 122 (with the literature cited there), and note especially Herrmann's association of the form with the vineyard-rich Fayum. In my view, however, the evidence for lease durations is equivocal; for the data, see Banaji 2007, table 12. *SB* XXVI 16722 (late VI) is the only example of indefinite duration (possibly) associated with the Apiones; note also LXVIII 4687 (441).

98. Cf. Beaucamp 2007, 282–83.

99. For the issue of volatility in the price of wine, see, most recently, Bang 2008, 156–73; cf. Erdkamp 2005, 67–70, and Rathbone 1997, 198–206. It should also be noted that Scheidel (2010, 456–57) argues for a significant increase in wages in the second half of the sixth century as a result of the Justinianic plague.

rent in cash? Might a tenant be forced to change the specie of his rent against his will? Might he be able to leave his tenancy under these (or other) circumstances? In other words, what was the place of coercion in labor relations?

Exploitation Evolving

The "lease of work." The issue of coercion by Oxyrhynchite landholders seems best explored through the evolution of labor deployment in the district. Jördens has noted, "There [in the Oxyrhynchite] the tenancy system had been partially abandoned quite early and had passed to direct exploitation by the landlord; the resulting gaps were initially closed through the engagement of paid agricultural workers."[100] The laborers to which Jördens is referring were governed by a novel form of contract, the so-called lease of work, or *misthōsis tōn ergōn,* which is attested from the late second century until the mid-fifth, primarily in the Oxyrhynchite.[101] Under these agreements, it was not the land itself that was leased but the "tenant's" work on the land. In exchange for this work, the tenant received a cash wage (*misthos*) and generally no (significant) share of the produce of the land itself.[102] Strikingly, all of these *misthōseis* concern the upkeep of vineyards or their supply of water, and one cannot help but think of the share site at Text I and the Hermopolite cropper contracts that have just been discussed—all the more so because the Hermopolite leases reference "manual viticultural labor" (*cheirikē ampelourgikē ergasia* or the like), an expression that echoes the object of most of the leases of work.[103] Jördens, who was not aware of the continued employment of sharecropping in the Byzantine Oxyrhynchite, in fact suggested that the lease of work supplanted share tenancy in that district.[104] There are, however, important differences between the agreements. None of the leases of work are much concerned with the vintage. "Tenants" may have to be present then,[105]

100. Jördens 1999, 138 (overstated).

101. There are a few Hermopolite examples. For this contract type, Jördens' research is fundamental; see *P.Heid.* V, ch. 4, as well as Jördens 1992b. Any generalizations based on the corpus of Oxyrhynchite leases of work must be tempered by its size; these texts number less than a dozen over a 250-year period.

102. Tenants might receive, e.g., a few jars of wine at the vintage; cf. *P.Oxy.* XLVII 3354.28. *P.Vind.Sal.* 8 is an exception; in this Hermopolite text, which combines a lease of vineyard work with a lease of land (an adjacent orchard), the lessee receives a third of the vintage (and intercropped produce) in addition to a wage. Note also *P.Col.* X 284, which seems to be structured similarly.

103. Cf., e.g., *P.Oxy.* XIV 1692.5: "all the manual viticultural work," τὰ ἀμπελουργικὰ χερικὰ [read χειρικὰ] ἔργα πάντα; XLVII 3354.5: τὰ χερικὰ [read χειρικὰ] ἀμπελουργικὰ ἔργα ἅπαντα.

104. *P.Heid.* V, pp. 227–29.

105. Cf., e.g., *P.Oxy.* XIV 1692.20–21.

they may have to make certain preparations for it,[106] and they may be required to take care of the fermenting must,[107] but nothing is said of harvesting the grapes, of treading or pressing, or of the procurement of the significant amount of labor that these activities required.[108] In contrast, Text I and the papyri from the dossier of Philoxenos suggest that (lessee) *ampelourgoi* were closely involved with production.[109]

Another noteworthy difference of the *misthōseis tōn ergōn* is the extent to which the work to be done is typically enumerated. It is true that Byzantine vineyard leases often include some reference to the tenant's work, even work standards,[110] but the precision of the leases of work is of another order.[111] Control is clearly the emphasis of these *misthōseis*, an emphasis that is especially apparent in those agreements in which the landlord disallows certain tasks, such as *P.Oxy.* XIV 1692, in which the object of the lease is τὰ ἀμπελουργικὰ χερικὰ [read χειρικὰ] ἔργα πάντα χωρὶς ξυλοτομίας, "all the manual vineyard work *except pruning.*" Presumably the landlord was unwilling to

---

106. Cf., e.g., *P.Oxy.* XLVII 3354.15–17: ποιησόμεθα τῆς ληνοῦ καὶ θυείου [pap.: θισυ] τὰ ἐνχρῄζοντα χαλάτρια καὶ τοῦ ὀργάνου τὰς κυλτίδας καὶ τὴν τῶν χωρούντων εἰς [pap.: ει] τὸν κατ' ἔτος οἶνον κούφων κομπασίαν ἀφ' οὗ τόπου μεταφέρεται, "We shall make the necessary matting for the treading vat and the press, and the *kyltides* [strainers?; cf. Kloppenborg 2006, 533] for the [pressing?] machine, and [we shall perform] the testing of the empty jars intended for each year's wine in the place from which they are transported." See also the next note.

107. Cf., e.g., *P.Oxy.* XIV 1631.16–18: ποιησόμεθα τὴν τῶν χωρούντων εἰς τὸν οἶνον κούφων κομπασίαν, καὶ ταῦτα λαβόντα τὸν οἶνον συνθήσομεν ἐν τῷ ἡλιαστηρίῳ καὶ ἐπαλείψομεν καὶ κεινήσομεν [read κινήσομεν] καὶ μεταδιαιράσομεν [read μεταδιεράσομεν] καὶ παραφυλάξομεν ἐς ὅσον ἐν ἡλιαστηρίῳ ἀπόκειται, "We shall perform the testing on the empty jars intended for the wine, and when these have been filled with wine, we shall put them in the drying room, and seal them, and move them, and filter the wine, and watch over the vessels as long as they are stored in the drying room."

108. In vineyards governed by leases of work, the vintage may have been addressed through *karpōnia* agreements (contracts for the purchase of an *unharvested* crop), for which cf. Rathbone 1991, 193–95.

109. Cf. also the penalty clause in some Arsinoite share leases, e.g., *SB* XVIII 13998.9–11: εἰ δὲ καὶ περιφρονήσω εἰς τὴν ἀπεργασίαν τῆς προειρημένης ἀμπέλου ἢ καταγνωσθῶ εἰς κλοπην τοῦ γενησομένου καρποῦ ἢ ζητήσω ἀναχωρῆσαι ἐξ αὐτῆς πρὸ συμπληρώσεως καὶ φθάσω τὸν καιρὸν τῆς τρύγης, "But if I am also negligent toward the upkeep of the aforesaid vineyard or am caught in the theft of the coming harvest or seek to withdraw from the vineyard before completion [of the contract term] and [in doing so] anticipate the time of the vintage." This suggests that the *ampelourgos'* presence at the vintage was deemed essential. It is not clear if the Apion estate assisted tenants with the procurement of labor for the vintage (such work might, e.g., be required as a labor due from residents in its *epoikia*). For this issue, cf. *P.Oxy.* LXXII, pp. 172–74 and refs.

110. *P.Hamb.* I 23.21–29 (Antinoopolite, 569) preserves an elaborate example. It should be noted, however, that such enumerations are not unique to leases of the late period; cf., e.g., *P.Soter.* 1, 2, and *SB* XXVI 16569.

111. See the useful tables in Rowlandson 1996, 325–26. A French vineyard labor contract described by Hoffman (1984, 312) sounds remarkably similar.

put this important procedure in the tenant's hands.[112] Thus a key difference between the lease of work and the share contract is the type of incentive employed. The former principally relies on negative reinforcement (i.e., monitoring) to check labor opportunism, while the latter provides a positive inducement for productivity in the form of a share of the yield.

Enapographoi geōrgoi. The last lease of work dates to 442, and it is the only exemplar that might pertain to the Apiones; it concerns the estate of Theodosius II's sister Arcadia that Strategius I might have been administering.[113] Even if this particular text does not have an Apion connection it would not be surprising if the family adopted the contract for their own holdings, given the influence of imperial models on "bureaucratic" aristocrats like the Apiones.[114] At the same time, however, another form of controlled labor was emerging: the *enapographos geōrgos* ("agriculturalist registered on a landowner's census") or *colonus adscripticius*—"a serf tied to the land by a hereditary bond," according to A. H. M. Jones's classic treatment of the status.[115] Though *enapographoi* have come to be seen as a hallmark of the Apion estate, it should be understood from the outset that the Egyptian countryside was not teeming with them. To date, *adscripticii* have been attested only in the Oxyrhynchite,[116] where they may have been limited to the estates of certain landholders. The papers of the Apiones furnish the largest number of references; next comes the dossier of Flavia Anastasia,[117] then scattered attestations in documents

112. Alternatively, if the "tenant" was due to receive a share of the yield (a rare circumstance; cf. *P.Vind.Sal.* 8), the landlord might be trying to discourage underpruning (which resulted in abundance and then depletion and is also known as "overcropping"). Overpruning is likewise problematic: the more vigorous canes that it yields create excessive shade; poorer fruit and increased pressure from pests result. For pruning abuses in the early modern period, see Pudney et al. 1998, 536, and Hoffman 1984, 312. Spreading fertilizer is also absent from the list of tasks in *P.Oxy.* 1692; when done improperly (i.e., overzealously in an attempt to bump up yields), it, too, can cause harm.

113. Cf. *P.Oxy.* L 3582 with Azzarello 2007, 253–54.

114. Cf. Banaji 1999, 203.

115. Jones 1958, 1 (still a valuable contribution despite the inappropriate language of feudalism). The literature on the colonate is voluminous and cannot be reviewed here; the papers in Lo Cascio 1997 (with the review article Scheidel 2000) are recommended as an introduction. Sirks 2008 is an important contribution of more recent date. Here, as elsewhere, my interest is in (regional) "practice on the ground" as it is revealed in the papyri, not so much in the legal sources that have hitherto driven the debate. The unsatisfying results of the approach through the law codes suggest that a new line of argument is needed. An assessment of the validity of the papyrological evidence outside of Egypt is beyond the scope of the present work.

116. There are one or two explainable exceptions; I do not accept the conclusions of Fikhman 1991, 10.

117. This is being prepared for publication by the author. Two published examples are *P.Oxy.* LXIX 4757 and *P.Select* 20.

concerning other *geouchoi* of various statuses as well as an institutional landlord (the church).[118] This limited distribution may simply be a function of the surviving evidence, but the geographical circumscription certainly is not. Even within the adscript estates, it is clear that certain document types that seem to be characteristic of the *enapographoi* might also involve those who lacked the status; thus, in *P.Oxy.* LXVII 4616, for example, Aurelius Biktor, an "unregistered" *geōrgos*, acknowledges receipt of a piece of irrigation equipment to the representative of Strategius II.[119] That the defining characteristic of the *enapographoi* was a fiscal one—namely, their landlord's guarantee of their tax[120]—is emphasized by the ubiquitous deeds of surety (*Bürgschaften*) of the late antique Oxyrhynchite.[121] Despite claims that a poll tax did not exist in Byzantine Egypt,[122] the principal tax that the landowner guaranteed for his *enapographoi* may well have been the *capitatio humana,* the *synteleia kephalēs* of the accounts of the *pronoētai.*[123] In support of this argument, it may be noted—this has never been observed—that no *epoikion* for which a complete pronoetic account is extant is attested both with *enapographoi* and without an account entry for the *synteleia kephalēs.*[124] If this pattern reflects reality, it

118. *P.Oxy.* XVI 1900 (the church), XXXIV 2724 (Fl. Kyria), XLIX 3512 (Fl. Ioannes *curialis*), LXIII 4398 (Fl. Dioskoros *scholastikos*), LXVIII 4697 (Fl. Phoibammon and Fl. Samuelios), LXIX 4755 (Fl. Ioannes, *vir gloriosissimus*); *P.Wash.Univ.* I 25 (Fl. Phoibammon *ho kai* Lamason); *SB* XVIII 13949 (Fl. Philodemos and Fl. Ioannes *scholastikoi*), 14006 (the church).

119. See also, e.g., *P.Oxy.* XVI 1899 (476); *PSI* I 60 (595, Apiones), III 179 (601 or 602, Apiones). Cf. *NJ* CXXVIII.14 (545): ἀλλὰ καὶ εἰ συμβαίη γεωργούς τινι προσήκοντας ἢ ἐναπογράφους ἰδίαν ἔχειν κτῆσιν . . . , "But if it should happen that the agriculuturalists attached to someone [i.e., a landowner] *or* registered in his census possess their own land . . ." *P.Oxy.* LXVIII 4688 (as well as some of the texts listed in its 7–8n.) could also be relevant, though Gonis may well be correct to argue that these represent "pre-adscript" phenomena.

120. Cf., e.g., Sirks 2008, 131–32.

121. We might also note that the very poor croppers in the Arsinoite (cf. p. 77 above) were not *enapographoi;* that is, being destitute was not enough to ensure the status. I suspect that an individual merely residing on a holding under the pagarchic authority of a *geouchos* (as opposed to a property *owned* by one of these grandees) would lack the status. In those few instances in which an *enapographos* dwells in a settlement said to be under a landholder's pagarchic authority (always a village; cf., e.g., *P.Oxy.* LXX 4787), I would suggest that he or she is living on the *geouchos*' property in that settlement.

122. E.g., P. M. Sijpesteijn 2007, 445–46.

123. So Sirks 2008; Sirks also makes the connection with the *synteleia kephalēs* (135 n.65). (In the same note, however, he misconstrues *phoros* as "tax.") *P.Oxy.* L 3584 also seems relevant for this question. In it, a *paroikos* states (ll. 6–7), τὸ τέλεσμα ὑπὲρ τῆς αἱρούσης μοι [κ]εφαλῆς εὐγνωμον[ῶ] καθ' ἕκαστον ἔτος, "I pay the head tax owed by me each year." Sarris (2006, 151–52) calls him "an adscript colonus *avant la lettre*" but seems not to recognize *caput* here.

124. *P.Oxy.* XIX 2243a.1–3 (with *P.Amh.* II 149) might be adduced as an exception, but I suspect that it is not a valid one, because the (fossilized) *lēmmata* in question are discharged by the *pronoētēs* to the account of one of his peers (2243a.61–65). I would suggest that we should seek the *synteleia* in this peer's account.

would indicate that adscript status was significantly limited even among the Apiones' labor force.[125] In any case, the fiscal nature of the status suggests that we should be wary of defining it by the work that those possessing it performed; as we shall see, the documents suggest some generalizations, but we should not be surprised by exceptions.

Prima facie, the evidence intimates that the *enapographoi* supplanted the tenants of the leases of work in the rural landscape of the Oxyrhynchite; at any rate, it seems probable that the *enapographoi* tended to be drawn from the same labor pool as the individuals who had worked under the *misthōseis*. We have already seen how the *enapographoi* have a similarly limited geographical range, and the chronology of the published documents concerning both groups is certainly suggestive.[126] Most telling, however, is the convergence of the work governed by the *misthōseis* and performed by the *enapographoi*. But there is an important difference: while viticultural work predominates in the leases of work, irrigation seems to have been a primary (or even the principal) charge of the *enapographoi*. In any case, the parallels indicate that wage labor should be our *point de départ* when we consider the vexed question of the nature of the Oxyrhynchite estates' labor relationships with the *adscripticii*.

One of the most common document types in which we encounter *enapographoi* is the receipt for a replacement part for irrigation machinery (sometimes referred to as a "waterwheel receipt"). Indeed, the recipients of these documents are typically *enapographoi*.[127] In the introduction, we saw that the receipts themselves are ambiguous regarding the relationship of the individual receiving the part with the estate, but there is good reason to think that they were usually employees. This is suggested not only by the leases of work but also by Jördens' hypothesis concerning the status of irrigators in Hermopolite share contracts and by my own analysis of the yield data for the share site from Text I (that a 3:1 division of the wine—i.e., one resulting from a lease in which an irrigator did not receive a share of the produce—is the best

125. Pace, e.g., Sarris 2006, 62.

126. The earliest Egyptian reference to an *enapographos* may occur in 440, if we accept the restoration of ἐναπόγραφος in *P.Mil.* II 64.5. Gonis has been critical of this; see *P.Oxy.* LXVIII 4697, 6–7n., but his objections do not seem compelling. (It is true that *enapographoi* of imperial estates have not been attested in the papyri, but cf. *CJ* XI.69.1.) If *P.Mil.* II 64 is rejected, *P.Oxy.* XXXIV 2724 (469) becomes the earliest papyrological reference; a papyrus among the Florence inedita (PSI inv. 839), however, could push the first attestation back into the fourth century (depending on the identity of its addressee, Flavius Macrobius; cf. *P.Oxy.* LXVI 4529, 3n., with Gonis 2002b, 91). Note also *ACO* II.1.2, p. 157, where *enapographos* appears in the *acta* of the Council of Chalcedon (451).

127. See n. 119 for some exceptions. As suggested in the introduction, the inherent fiscality of adscript status may well skew the documentary record.

fit).[128] Thus I would propose that the teams working in the Apion vineyards that yielded in-kind rents typically would have consisted of a tenant (or tenants) and an estate employee (or employees); the former, typically an *ampelourgos,* would have concerned himself with the vines and vintage, the latter with irrigation. The language of Text I (e.g., πρὸς *personal name* [in the accusative] ἀμπελουργὸν *personal name* [in the genitive] γεωργοῦ) may indicate that the estate considered the vineyard to be assigned to the irrigator, who might also have monitored estate interests.[129]

Several machinery receipts would seem, however, to stand against the argument that the *enapographoi* involved with irrigation were employees. These texts specify a rent, and an *enapographos* is involved in one case.[130] We have already seen that *enapographoi* might pay rents,[131] so this aspect of the texts, while admittedly exceptional in the documentary record, should not be considered controversial. More to the point, all of the irrigation receipts in question date from the fifth century, which suggests that the combination of irrigation work and a lease may have died out in the Oxyrhynchite by the sixth century; none of the Hermopolite contracts in this form were drafted after 510.[132] This may indicate the deterioration of the fortunes of those involved in irrigation work.

Such deterioration may well be signaled by another document type in which the Oxyrhynchite *enapographoi* commonly appear: the *prochreia* agreement. Once again, these documents suggest the geographic distribution of the lease of work: they are attested far more often in the Hermopolite and Oxyrhynchite than in other nomes.[133] Moreover, a significant number of them concern irrigation work or work on a *mēchanē.*[134] Generally speaking,

---

128. Cf. also *SPP* III² 86, which concerns an Apion property and is quoted above (p. 63).

129. Cf. Sirks 2008, 135–36. Sirks seems to have been led to suggest the use of *enapographoi* as managers by the presence of women in irrigation machinery receipts (cf., e.g., *P.Oxy.* LXX 4797). But if children were capable irrigators (cf. p. 51 n. 67 above), women were certainly equal to the task.

130. See *P.Mil.* II 64.12 (imperial); *P.Oxy.* XVI 1899.20–21 (Fl. Alexandros), 1982.20–21 (Apiones, involves an *enapographos*).

131. In *P.Oxy.* XVI 1896. Note also, e.g., *P.Oxy.* XXVII 2478, a surety for an *enapographos* that employs the same verb for payment (l. 19: *eugnōmoneō*) as the irrigation receipts, as well as *P.Oxy.* LXVII 4615.

132. For the irrigation share lease elsewhere in Egypt, cf. n. 83 above.

133. In the Oxyrhynchite, the recipients are usually *enapographoi* (cf. Jördens 1999, 140–41); this is always the case in an agricultural context. If one includes Coptic texts, the Theban region is also well represented.

134. Irrigation work: Cf. *SB* VI 9284, *P.Lond.* III 1037 (pp. 275–76), and *P.Flor.* I 70, all from the Hermopolite. *Mēchanai:* Cf. *SB* XXII 15362 and *P.Oxy.* I 194 (to be published by the author and Amin Benaissa). See also *P.Iand.* III 48, whose *enapographos* is linked to a *mēchanē* in an irrigation machinery receipt (*P.Oxy.* I 137). Jördens (1997, 81) notes, "Besides general services, irrigation

prochreia is a wage advance in the guise of a loan,[135] and Jördens has indicated that it was a vehicle by which *geouchoi* increased the ties of dependency.[136] This may well be the case, but *prochreia* should also be reckoned as a possible *symptom* of dependency—in other words, as an influx of capital, often to be consumed during the completion of a task, for those lacking adequate resources. Though it is from the Theban region, *O.Vind.Copt.* 28 seems helpful for understanding what must have been a common Oxyrhynchite situation. In this text, a certain Isak declares that he will sow, work, and irrigate a certain plot of land in return for the *semissis* (half-*solidus*) given to him to pay his tax. This is, in its essence, what the *enapographoi* had agreed to do, and I am struck by the fact that Ioseph, the lender, states, "If you come to me on account of anything else, I shall provide it for you."[137] Ioseph is no doubt referring to resources needed to complete Isak's agricultural tasks, and this suggests that the *prochreia* advances given to the *enapographoi* were often aimed at the work assigned to them.[138]

Why? For the causes of dependency, we need look no further than incomes. Sirks has stated that the going wage for the *enapographoi* was "some food and some pocket money."[139] Even with respect to the "normative" *adscripticius*—say, an irrigator—this is an understatement, though the additional resources likely did nothing to lessen his dependency on the *geouchos*.[140] The *enapographos* and his (similarly adscript) family no doubt received shelter in the *epoikia* that are ubiquitous in the late Oxyrhynchite documentation,[141] and his landlord may have provided him with a parcel of land for subsistence,

work was most often arranged." Also in the work realm of the *misthōseis tōn ergōn* is *P.Amh.* II 149, a *prochreia* agreement with an *enapographos pōmaritēs* ("registered orchardist"). The Coptic documentation suggests usage with flax cultivation (cf., e.g., *P.Mon.Epiph.* 85; *O.CrumVC* 29), for which water was crucial.

135. Cf. *P.Heid.* V, chapter 7.

136. Cf., e.g., Jördens 1997, 81: "Since this one [the giver of *prochreia*] is typically the landlord, we can see one of the ways into the manifestly feudal structures of late Greek Middle Egypt." Dependency is also emphasized by Mazza (2007, 441–43), who likewise highlights the employment of *prochreia* in leases, but most *enapographoi* (including, e.g., the one in *P.Oxy.* 192, cited on her p. 442) were not working under such agreements.

137. ⲉⲕϣⲁⲛⲉⲓ ⲛⲁⲓ ϩⲁ ϩⲱⲃ ⲛⲓⲙ ⲉⲧⲉⲣⲟⲥ ⲉⲓⲛⲁⲧⲟⲩϣϥ ⲛⲁⲕ.

138. Contra Banaji 1999, 210–11.

139. Sirks 1993, 363.

140. Somewhat less dependent would have been nonadscript wage laborers, but much in my description here is relevant for them as well.

141. The argument for labor tenancy on Byzantine Egyptian estates is, in my opinion, one of Banaji's most important contributions to the scholarship; cf., e.g., Banaji 1999. Estate accommodation is probably attested in *P.Oxy.* XVIII 2197.74, 143, 150, 152–56. For the barracks in the *epoikia*, which were presumably occupied by single men, cf., e.g., *P.Oxy.* XVI 1917.56.

perhaps at a subsidized rent.[142] In return for this, he may have been liable for some labor service (beyond his ordinary responsibilities),[143] and he was deprived of the freedom to leave the estate without authorization.[144] This restriction, as well as obvious estate attempts to make the *epoikion* a community unto itself,[145] was at least in part intended to limit the horizons of the *enapographoi.*[146]

Our irrigator's partner in the vineyard, the share tenant, would have been in a better position. On an average vineyard of 2.25 *arourai* that yielded 33.4 hectoliters per hectare, his share would have been 948.72 *sextarii,* perhaps valued at 3.95 *solidi* "in the company store"[147] and worth up to 6.59 *solidi* at more favorable prices.[148] Although this is a gross yield, before expenses and extra charges and any deductions for spoilage,[149] its value could be compared to the salary (with food allowance) of an Apion *pronoētēs.*[150]

---

142. Often such plots are given in exchange for labor (i.e., without a rent in cash or kind), and, accordingly, we would not expect to see them in estate accounts. But entries in the pronoetic accounts that at "normal" tariffs equate to parcels that would have been insufficient for subsistence suggest the possibility that these were allotted to employees at a subsidized rent (for similar subsidies in modern Egypt, cf. Nour ed-Din 1898, 5–6). Some figures to consider:

1. Bagnall (1993a, 70) puts *household* subsistence at 10 *arourai.*
2. Banaji (2007, 192) suggests that the rent of a perennially irrigated *aroura* on the Apion estate was 12 *keratia* or 7 *artabai.* This is probably too high for the kind of plot that labor tenants would have received, but it is a ceiling.
3. In the account *P.Oxy.* LV 3804, e.g., 63 percent of the holdings among the *lēmmata* concerning agricultural land fall beneath Bagnall's subsistence level when converted using Banaji's figures. With a 50 percent rent subsidy (the figure attested in Nour ed-Din), this number drops to 47 percent. At 67 percent, it is 37 percent; at 75 percent, it is 24 percent. These are crude calculations; the issue is a significant one, and a more careful assessment than is permitted here would be most welcome.

143. *P.Wash.Univ.* II 102 (+ *BL* XII 289) is the locus classicus. Cf. also *P.Oxy.* LXXII 4926–29 (but note that 4927 and 4929 concern villagers).

144. In theory, at least; cf. Sirks 2008, 136–38. Based on the large number of sureties for *enapographoi* that are extant, Sirks notes, "Flight was evidently a problem" (137).

145. Churches were a typical feature of the *epoikia;* cf., e.g., *P.Oxy.* XVI 1910.2–4, 1912.115–19, XVIII 2195.84–88, 187, XIX 2243a.75–78, LV 3804.144–48. Note also the network analysis in Ruffini 2008, which suggests few horizontal connections for the Apiones' *epoikia.*

146. Cf. Schaffner 1995, 247, summarizing the research of A. R. Garcia concerning sugarcane plantations in northeastern Brazil: "[E]mployers . . . sought to 'limit' the 'mental and social horizons' . . . of their permanent workers by contracting them to reside on the plantations, providing plantation churches, stores, bars and other substitutes for facilities in town, and in some cases explicitly forbidding workers to leave the premises. In fact residence on the farm, rather than permanence of employment, was the salient feature giving these workers their name, *moradores* (literally, residers). Their limited horizons . . . facilitated their domination by employers."

147. I.e., according to the internal estate tariff of 240 *sextarii* per *solidus* (cf. n. 54 above).

148. My informed hypothesis for a price ceiling is 144 *sextarii* per *solidus.*

149. Cf. p. 77 n. 90 above.

150. Banaji (2007, 236) reports a *pronoētēs'* salary as 3.79 *solidi,* but this is missing the wine component (at the least); also note p. 135 below. It will be observed that there are arrears at the Text I share

The share tenants' peers at the fixed-rate sites received a somewhat larger share of the yield, but this must had to be divided between more workers, resulting in a significantly thinner margin. The worst yield value at the *epoikion* Nekontheos is 1.52–2.53 *solidi* per man, unless the irrigator(s) also received a share,[151] in which case it is .67–.81 *solidus*. One might think to compare this to the three *solidi* that the pious *thermodotēs* received from his employer as an annual wage in the vita of St. John the Almoner,[152] and, in any case, it comes as no surprise that a *koinon* of *ampelourgoi* from an otherwise unattested *epoikion* petitions against the Nekonthite rent rate (150 *dipla/ar.*) in *SB* XVI 12554.[153] In the context of limited incomes, the attractions of fiscal guarantees, of adscript status and its attendant benefits, and of the protections of the circumscribed world of the *epoikion* become clearer. We can understand now why the fixed-rate tenants at the *epoikion* Leontos might have leapt into the adscripticiate:[154] it seemed better than the alternative.[155] Yet, at the same time, we must remember that the extent of the adscripticiate on the Apion estate remains undetermined; as we have seen, it may have been confined to certain *epoikia*, and even if it was a feature of every estate settlement, it was largely limited to certain kinds of labor (in particular, irrigation work). Tenancy, which the sources indicate was the principal means by which the estate exploited its

---

site: shortfalls from the prior (i.e., fourteenth) indiction are attested for all but one of the seven tenants. The deficits range from 12 to 37 *dipla;* the mean is about 19 *dipla,* a bit over 5 percent of the mean current liability. At the end of the fifteenth indiction, all seven tenants come to owe arrears, but in five out of seven cases, these have gone down by an average of about 11 *dipla.* Of the increases, one is by only a single *diploun;* the tenant that owed no arrears after the fourteenth indiction, however, owed 20.25 *dipla* after the fifteenth. Since Text I is a snapshot—and we have no way of knowing the tenants' complete "portfolios" and transactions—it is hard to assess this data (but see n. 90 above). The case of Soterichos (on whom cf. Kehoe 1992, 141–49) comes to mind as a cautionary exemplar.

151. As noted above (p. 77 n.88), the labor structure of the fixed-rate vineyards is less clear.

152. Prol. 125–27 (p. 345, Festugière-Rydén): εἶς οὖν τῶν συγκαθημένων εἶχεν ἐκεῖ παριστάμενον μίσθιον, θερμοδότην αὐτοῦ ὄντα, λαμβάνοντα τρία νομίσματα κατὰ ἐνιαυτὸν καὶ μόνον, ἔχοντα γυναῖκα καὶ δύο παιδία, "One of those sitting in the group had an employee standing by there, a man who brought his hot water, who received three *solidi* annually and no more, although he had a wife and two children." If anything, Leontios is understating the *thermodotēs*' salary to highlight his generosity.

153. This text, in which the petitioners seek to pay 100 *dipla* per *aroura,* has no obvious link to the Apiones.

In Text I, there are shortfalls at the two fixed-rate sites. The two work groups at Phankona owe arrears from the previous (fourteenth) indiction, both of them about 5 percent of the quantity presently due. At Nekonthis, only one group is still liable for the past indiction (7 percent of the total due); in the indiction covered by the account, however, all of the Nekonthite vineyards have fallen short on their rents, the one that already owed arrears by a whopping 311 *dipla.* Incidentally, the members of this team would have received the largest shares of must in a normative year. At Phankona in the fifteenth indiction, there is a slight reduction in arrears.

154. *P.Oxy.* XVI 1896.

155. Cf., most famously, Matthew 20:1–16.

vineyards, attests *enapographoi* only exceptionally,[156] and it may be that *adscripticii* were only able to pay rents in kind.[157] If this was indeed the case, we can eliminate those in the pronoetic accounts who pay cash land rents—a huge mass of people—from the ranks of potential *adscripticii*.

If, however, one wishes to grant the premise that "the mainstay of the Apion family's agricultural workforce was made up of *coloni adscripticii*,"[158] the source(s) of such widespread deterioration in the status of labor must be sought. In the pursuit of comparanda, the (in)famous "second serfdom" of northeastern Europe comes to mind most immediately, all the more so given that Le Roy Ladurie suggested comparing it to "the rise of the colonate."[159] Among the causes that have been proposed for the second serfdom are demographic decline (in part, but not exclusively, from plague) and demand from external markets.[160] Scheidel has recently argued that the Justinianic plague had an impact on the Egyptian labor market, though his research documents not more robust labor controls but, rather, an increase in wages (as in England under the Black Death, the classic "Malthusian" scenario).[161] The validity of Scheidel's data set aside—the monetary phenomena of Byzantine Egypt present considerable problems of interpretation—there are other factors to consider, such as the (im)mobility of Egyptian labor[162] and, more broadly, whether the institutions prevalent in late antique Egypt and medieval England lead us to expect comparable developments.[163] Mazza, at any rate, has emphasized labor scarcity, noting, "The sureties and warranties present in nearly all of the documents of the archive attest to this sort of obsession of the owners, who feared losing farmers [Italian *coloni*] and thus the failure to

156. Cf. p. 83 above.

157. Cf. Sirks 2008, 133.

158. Sarris (2006, 62) sees this as irrefutable.

159. Le Roy Ladurie 1985, 104 (albeit in late Roman Gaul).

160. Brenner (1985, 37) summarizes, "Serfdom began its rise in the east . . . in the period of late medieval demographic decline; it was consolidated during the trans-European increase in population of the sixteenth and seventeenth centuries [hence increased demand]; and it was further sharpened at the time of the demographic disasters of the later seventeenth century." The persistence of serfdom despite these population shifts (alongside differences between developments in the West and those in the East) leads the (highly controversial) Brenner to reject demographic explanations. Cf. also, e.g., Dyer 2004, 111 (critical of the decline argument), and Anderson 1974, 196–97 (critical of the demand argument).

161. Scheidel 2010, 456–57.

162. Cf. J. Abu-Lughod 1989, 238 (concerning medieval Egypt, but meriting our consideration for the earlier period). I note that the movement of labor was already being constrained before the Justinianic plague; cf., e.g., *P.Wash.Univ.* I 25 (530).

163. Cf. Borsch 2005, attributing the divergent impacts of plague in Mamluk Egypt and medieval England to institutional differences.

appropriate the *reditus* [return, income] necessary to run their estate."[164] The matter, in short, would benefit from further scrutiny, though it is not an issue for which the data from the Apion dossier are well suited.[165] The question of market demand, in contrast, is quite apposite to any study of estate viticulture; indeed, as we saw in chapter 1, viticulture is the only sector in which significant market engagement could have occurred. To begin to examine this proposition, which has received so much emphasis in recent scholarship, we must illuminate the movement of wine through the estate. This is the focus of the next chapter.

164. Mazza 2001, 160.

165. Egyptian papyri are all but silent about the pandemic, though in the case of the Apion estate, the majority of the documents date to the period after its arrival (see p. 21 above). Note, further, Hickey 2008 (on the change in estate target receipts and taxes over the second half of the sixth century), especially 99 n. 58 and refs. (on the plague).

Chapter 4

The Flow of Wine through the Estate

In chapter 1, I demonstrated that viticulture is pivotal for any analysis of the Apion estate's posture toward the market. In chapter 3, I noted that one might argue that developments in the estate's relationships with labor were a reaction to demands from the market. The aim of the present chapter is to begin the evaluation of the relationship between estate viticulture and the market by examining the movement of wine into and out of the *oikos.*

Almost fifty years ago, Rémondon observed, "Regarding the estate of the Apions, it was autarkic (the Apions were in the habit of sending the wine from their own vineyards to the court in Constantinople when they where in residence there)."[1] Though such a distinguished assessment should carry weight,[2] it need not be accepted at face value, for we possess ample evidence for the disposition of estate wine in the form of the annual accounts of the *oinocheiristēs,* or wine steward. Three specimens of this text especially deserve our attention because they are so well preserved: *P.Oxy.* XXVII 2480 (580/81),[3]

1. Rémondon 1964, 311.

2. It was rejected, however, by Rémondon's own disciple Gascou; see Gascou 1985, 10 n. 35.

3. For this date (pace ed. pr. [565/66]), see the discussion that follows. Seven fragments of this text (in addition to the exiguous remains of col. 1) are unedited. *P.Oxy.* XIX 2244 fr. 1 does not belong to 2480. It should be dated to the period after *P.Princ.* II 96 (and *PSI* VIII 955; for the chronology of these texts, see the next note); 2244 fr. 1.3 reveals that the Philoxenos of 96.4 (and 955 ined.) has retired.

P.Oxy. LVIII 3960 (621), and *PSI* VIII 953 (567/68).[4] But some words concerning the *oinocheiristēs* himself are necessary before we delve into the rich details of the accounts.

The *Oinocheiristēs*

In his introduction to *P.Oxy.* LVIII 3960, John Rea has written that the *oinocheiristēs* "seems to have been in charge of the Apion estate rather in the way that the general stewards (προνοηταί) were in charge of a group of Apion holdings which were near enough to one another to be managed jointly";[5] that is, Rea believes that the estate had several *oinocheiristai.* He adduces two texts, *P.Oxy.* XVI 1916r and 2044, in support of his argument, noting that these accounts have three and four of the sites, respectively, that appear in *P.Oxy.* LVIII 3960; because of this, he suggests that all of the places in question were located close enough to be administered conveniently. Of course, such reasoning is only valid if the two supporting texts concern limited areas. The former surely does not; it is an account of the *dikaion* Kephala (i.e., property still registered in the cadastre in Kephalas' name),[6] and since the *epoikion* Nekontheos (in the former Lower Toparchy), the *kōmē* Tampeti (seventh *pagus*), the *kōmē* Adaiou (fourth *pagus*), and the *epoikion* Kotyleeiou (Upper Toparchy) all appear,[7] it is clear that Kephalas' former holdings were dispersed throughout the Oxyrhynchite. The case for *P.Oxy.* XVI 2044 is less certain, but sites from the former seventh *pagus* (*epoikion* Leonidou) and the former Eastern Toparchy (*kōmē* Adaiou) appear, and I would suggest that it is in fact an estate-wide account.[8] *P.Oxy.* LVIII 3960 itself has sites from the former seventh *pagus*

4. The fragments published as *PSI* VIII 956 belong to 953, as is implicit in Gascou 1972, 70 n. 3. But note already *PSI* VIII, p. 150, and cf. lines 7 and 53 of 953 with lines 29 and 23 of 956, respectively. The fragments edited as *PSI* VIII 955 (+ *SB* XX 15181) belong to another account. They date to a fourth indiction—probably 570/71, since the Philoxenos *meizoteros* in an unedited fragment appears in *P.Princ.* II 96 (566/67, not 551/52; for discussion, see *P.Oxy.* LXX 4787, 5n., and cf. *PSI* III 191 with, e.g., line 74 of 96). Roberta Mazza and I will be publishing this new fragment and four others in Florence related to 953 and 955. Sixty-four additional Florentine fragments of stewards' accounts remain unedited. PSI inv. 1315, to be published by Isabella Andorlini, joins to *PSI* 953.54. PSI inv. 4058–4120, to be edited by Andorlini and Federico Morelli, derive from various accounts; first, second, fourth, and fifteenth indictions are certainly mentioned in the papyri of this group. (I thank Guido Bastianini and Isabella Andorlini for this information and for their hospitality while I was working in the Florence collection.)

5. *P.Oxy.* LVIII 3960, p. 118.

6. For *dikaion,* see pp. 58–60 above.

7. For these sites (and others mentioned in this discussion), see appendix B, s.vv.

8. For further discussion of this text, see p. 96 n. 20 below.

(*epoikion* Leonidou), the fourth *pagus* (*kōmē* Adaiou), the third *pagus* (*kōmē* Episemou), and the former Upper Toparchy (*epoikion* Kotyleeiou), as well as Oxyrhynchus itself. The *epoikion* Orthoniou is also mentioned, and if that settlement was indeed located in the former Lower Toparchy, it would seem rather unlikely that another steward (conceivably based in Herakleopolis) was charged with the administration of the northern portion of the Oxyrhynchite. In short, there would seem to have been only one *oinocheiristēs* on the estate. Confirmation is provided by 3960.20–21, which documents a distribution "to the holy churches and hostels and *martyria* of the city and in the country," εἰς τὰς ἁγίας ἐκκλησίας καὶ ξενοδοχεῖα καὶ μαρτύρια τῆς πόλεως καὶ ἐν ἄγροις. Here the *oinocheiristēs* is concerned with Oxyrhynchus and its *enoria* (*territorium*), which corresponded to the entire nome.[9]

Table 4.1 lists all attestations of *oinocheiristai* known to me. Though management of wine stocks was certainly required well before late antiquity, the term *oinocheiristēs* does not appear until the fourth century (4.1.20). Its use was apparently confined to the Oxyrhynchite and the Arsinoite; presumably, the function had a different title in other places (or its duties were handled by several individuals). More than two-fifths of the attestations have firm links to the Apiones (4.1.1, 3, 7, 11, 14–15, 17–18, 21), while further references concern other private landowners (e.g., the *comes* Ioannes in 4.1.16) and the church (e.g., 4.1.9).[10] None of the texts come from a certain public milieu (both 4.1.8 and 4.1.10 are ambiguous).[11]

It will be observed that the tenures of the Apion stewards Serenos (4.1.18) and Phib (4.1.14) overlap if one accepts the datings current in the scholarship—more precisely, the date of *P.Oxy.* XXVII 2480, since the *PSI* dates are absolutely secure.[12] This of course speaks against the argument for a single *oinocheiristēs*. There are, however, strong indications that 2480 has been misdated. For one, the expenditures in both stewards' accounts (Phib: 2480; Serenos: *PSI* VIII 953) are centered on Oxyrhynchus. At the least, this is an obstacle to a territorial organization for the office of *oinocheiristēs*, and the fact that no thematic division of labor is apparent in the accounts in question spurs one to consider a different date for the less chronologically certain text

9. *W.Chr.* 281.33, 45–46, makes the connection between *enoria* and *agros* clear. For *enoria* in the sixth century, cf., e.g., *SB* XXIV 16312.10–11.

10. For Count Ioannes, cf. Hagedorn and Kramer 2004, 158–61.

11. 4.1.8: A *castrensis* and *oikonomos* (steward) might be a public figure, but the nature of the *castrensis* in *P.Turner* 51, e.g., gives pause. 4.1.10: That the text has *oino(cheiristēs)* instead of, say, *oino(pōlēs)*, "wineseller," is far from certain.

12. Cf. especially the dates of *PSI* III 191–93 with *P.Oxy.* XXVII 2480.26–27, 29–30, 33–34.

TABLE 4.1. *Oinocheiristai*

Name	Attestations (Provenance)	Date(s)	Notes
1. Athanasios	Text I (Oxy.)	551/52	
2. Biktor/Ouiktor (I)	—*P.Oxy.* LXVII 4622.1? —P.Wash.Univ. inv. 251.1 (Oxy.)[a] —*SB* XVI 12608.1 (Oxy.) —P.Oxy. *inedita*	4622: V/VI 12608: 11.vii.511 *inedita:* 504–11[b]	—Not to be identified with the *oinopratēs* in *P.Oxy.* LXVII 4621, *SB* XXII 15528–30[c]
3. Biktor (II)	*P.Amh.* II 183 *descr.* (Oxy.)[d]	VI (≤590; ca. 540?)	—This individual's heirs are probably referenced in *P.Oxy.* XIX 2243a.37 (28.x–26.xi.590), where he is styled *apo oinocheiristōn.*[e]
4. Dorotheos	*P.Wash.Univ.* II 99.2 (Oxy.)	V–VI	
5. Georgios	*Stud.Pal.* XX 196.1 (Ars.?)	V–VI	—XX 196 = *SB* I 5297/*Stud.Pal.* VIII 1160
6. Ioannes (I)	*SB* XX 14073.1 (provenance?)	VI–VII	—same individual as 7?
7. Ioannes (II)	*P.Oxy.* LVIII 3960.3	621	—same individual as 6?
8. Markos	*BGU* II 692.1 (Ars.)	IV–VIII	—II 692 = *Stud.Pal.* VIII 1149
9. Menas (I)	*P.Oxy.* XVI 1951.2	V (485?)	—An employee of the church, so not the same individual as 10
10. Menas (II)	*SB* XX 14644.1 (Oxy.)	VI	—XX 14644 = reed. of *P.Wash.Univ.* I 48 —Ed.: *oino(cheiristēs) poleōs.*
11. Pamouthis	*P.Oxy.* XVI 2032.62, 64	540–41 (cf. *BL* VI 105)	
12. Petros (I)	*BGU* II 693.1 (Ars.?)	IV–VII	—II 693 = *Stud.Pal.* VIII 1150 —same individual as 13?
13. Petros (II)	*SB* VI 9398.2 (Ars.)	VI/VII	—same individual as 12?

(continues)

TABLE 4.1—*(Continued)*

Name	Attestations (Provenance)	Date(s)	Notes
14. Phib	—*P.Oxy.* XVI 2044.1 —*P.Oxy.* XXVII 2480.121, v.1 (Oxy.)	2480: 565–566[f] 2044: 25.vii–23.viii.564 (cf. *BL* XII 144; the text almost certainly predates 2480)	
15. Philoxenos	—*P.Oxy.* X 1326.2 —*SB* XXII 15320.2 (Oxy.)	Second half V[g]	
16. Phoibammon (I)	—*P.Oxy.* I 141.1, 6 —*P.Oxy.* LXVIII 4699.1 —*PSI* VIII 957.1, 7 (Oxy.)	141: 19.xii.503 4699: 23.1.504 957: 29.i.504 (*BL* XI 248–49)	—I 141 = *Stud.Pal.* VIII 1155
17. Phoibammon (II)	—P.Iand. inv. 473[h] (Oxy.) —*P.Oxy.* I 150.1 —*P.Oxy.* XVI 2012.1 —*SB* XXII 15368.1 (Oxy.)	2012: 30.ix.590 150: 7.x.590 15368: 27.x.590	—I 150 = *Stud.Pal.* III 283 (HGv incorrectly ascribes to Herakleopolite)
18. Serenos	—*PSI* III 191.1 (Oxy.) —*PSI* III 192.1 (Oxy.) —*PSI* III 193.1 (Oxy.) —*PSI* VIII 953.1 (Oxy.)	193: 28.ix.565 (*BL* VII 233) 191: 5.x.565 (*BL* VII 233) 192: 27.x.565 (*BL* VII 233) 953: 567–68 (*BL* V 125)	
19.]gios	*Stud.Pal.* VIII 997.1 (Ars.?)	VI	—same individual as 5?
20. ?	*P.Oxy.* XIV 1752.1	26.iv.379 (*BL* VIII 249)	
21. ?	*P.Oxy.* VIII 1131.1	mid VI (*BL* X 141)	—a person in this text also appears in *P.Oxy.* XVI 2032 (cf. 11 above.)

[a]The text will appear in *P.Wash.Univ.* III.

[b]Information kindly furnished by Nick Gonis.

[c]Cf. Papaconstantinou 2001, 280. (The objection to the identification in *P.Oxy.* LXVII 4622, 1n., seems over rigid; for the probable involvement of the *oinocheiristēs* in the wine trade, see p. 141.)

[d]A *descriptum* to be published by the author elsewhere.

[e]Given the fixed/fossilized nature of the *lēmmata* in accounts like 2243a, a gap of 50 years between that text and the Amherst papyrus would not be problematic.

[f]So ed. pr., incorrectly; see pp. 95–98.

[g]So *APF* 40 (1994) 119, regarding 15320. "Final quarter of the fifth century" is more accurate (so Nick Gonis, who will be publishing the dossier in full with Amin Benaissa).

[h]Described in Azzarello 2007, 258–59.

Addenda: The *oinocheiristēs* Chairemon (*P.Oxy.* LXIII 4373 [26.vi.364]) is absent from the list. *P.Laur.* IV 178 (Oxy., 17.ix.557/572) furnishes another attestation of Serenos (table 4.1.18). In 4.1.20, the name of the steward is Neilos.

(i.e., 2480). Accounting procedures in *PSI* VIII 953 and *P.Oxy.* XXVII 2480 are also somewhat at variance;[13] this admittedly could simply reflect quirks of the stewards, but the estate's record-keeping practice may also have changed over time.

The date of Easter in *P.Oxy.* XXVII 2480 creates more doubts. If one accepts the fiscal year currently assigned to the account, the references to Easter (ll. 84, 96) must correspond to a Julian date of 28 March 566, Pharmouthi 2 on the Egyptian calendar. Yet there is a rough chronological order to the entries in 2480 under Pharmouthi, according to which we would expect Easter to appear at or near the beginning. It does not. The entries run as follows: 31 March, 10 April, 11, 11, 11, 12, 12, 13–15, 13–15, 13–15, 13, 15, Easter, 14–15, 13, 14–15, 15, 15, 16, 16, Easter, 9, 21. In light of this sequence, an Easter date in mid-April seems preferable, but one must go back to 491 (when the feast fell on 14 April) to find one. The next best options are 521 (11 April), 551 (9 April), 581 (6 April), and 596 (22 April), but for prosopographical reasons,[14] only 551 and 581 are probable alternatives. The mention of the estate prison (ἐν τῇ φυλακῇ τοῦ ἐνδόξου οἴκου, l. 84) rules out 551, since the Apiones were still using the city prison in 552;[15] moreover, we would expect the entry for Pharmouthi 14 (= 9 April; l. 96) to be tagged ἐν τῇ ἁγίᾳ Ἀναστάσει ("on the Holy Resurrection").[16]

This leaves the original 566 or 581, and the latter date seems preferable for several reasons. For one, the steward Phib's account records a distribution of wine to a Papnouthis *epikeimenos* (overseer; ll. 20–21). This same individual probably reappears in *P.Oxy.* XVIII 2197.1, for which a date in the 580s has recently been suggested on the basis of another (i.e., independent) prosopographical link.[17] A later date for *P.Oxy.* XXVII 2480 may also help explain a quantitative discrepancy. In 2480, the steward's *hypodochē* ("reserve") for the fourteenth indiction is 19,321 *dipla,* while in *PSI* VIII 953, from just two indictions later—if we accept 2480's original date—the steward expends over 36,700 *dipla.* These *dipla* were the same size, so it is not a question of different measures,[18] nor does it seem probable that 2480 attests the typical *hypodochē* and

13. Cf., e.g., p. 105 below.

14. Cf. already Skeat in the introduction to 2480, p. 182: "P.Princ. ii 96, which includes some of the same persons as 2480, and which was independently dated by Kase to 551–2 or 566–7."

15. I thank Nick Gonis for this information, which appears in an unpublished surety.

16. A similar impediment applies to the indictions 490/91 and 520/21.

17. See *P.Oxy.* LXX 4792, 10n. Papnouthis also appears in *P.Oxy.* LV 3805.35, an example of *reuse* dated to 566 or later. The Theodoros *meizoteros* in *P.Oxy.* XXVII 2480 (l. 3, e.g.) probably also reappears in *P.Oxy.* XVI 1861 (dated VI–VII) and perhaps in *P.Oxy.* LIX 4006 (dated VI–VII). For other *meizoteroi* with this name, see 4006, 11n.

18. See p. 124 below.

Fig. 5. From the annual account of Phib *oinocheiristēs* (*P.Oxy.* XXVII 2480, cols. 1–2). (© The British Library Board.)

that the difference in *PSI* VIII 953 was made up through purchase; in *P.Oxy.* LVIII 3960, the steward's reserve, even after the Sasanian invasion and occupation, stands at 33,298 *knidia.*[19] Though one might attribute the low figure in 2480 to a problem in the vineyards,[20] a date of 581 allows for a more com-

19. These *knidia* have the same capacity as the *dipla* in the earlier texts (cf. p. 99 n. 35). Approximately 7.5 percent of 3960's wine was acquired outside of the estate.

20. *P.Oxy.* XVI 2044 is an account of *oxos* (vinegar) "excluded" (ἀποκλεισθέντος) from Phib's *hypodochē* in a twelfth indiction. The account is associated with Mesore (the usual month of the vintage), and this, along with the large quantities of *oxos* involved (8,143 *dipla* in total), suggests that the text concerns product that is at least a year old. *Oxos* would have been more common in such wine, as the wine vessels were not airtight, and this allowed aerobic fermentation (and microbial

pelling hypothesis: namely, that it reflects a division within the Oxyrhynchite *oikos* between the heirs of Apion II.[21] This need not have been an actual apportionment of property—indeed, there would have been compelling economic (and fiscal) reasons to keep the *oikos* intact[22]—but a division of income and expenses, that is, an accountant's apportionment.[23] In favor of this hypothesis, it may be noted that the *hypodochē* in 2480 is almost exactly half of the expenditure in *PSI* VIII 953, which may reflect partition of the *lēmmata* and *analōmata* into two equal shares.[24] Finally, *P.Oxy.* XXVIII 2480 furnishes the earliest fixed reference to the estate *antigeouchos*, or landlord's representative; it seems much more sensible to ascribe the creation of this position (and a new layer of management) to the uncertain period following Apion II's death.[25]

infection) to occur. Year-old wine, moreover, would have had to endure a summer of heat (cf. Rathbone 1991, 258). At the same time, however, it is curious that the sources of these jars of *oxos* are noted in the text, for it seems rather unlikely that their producers would have been held responsible for any souring that occurred so long after the vintage. Such "tracking" suggests recent production, since estate sharecroppers seem to have been held responsible for the quality of their wine for a period (see p. 77 n. 90 above), while the typical Oxyrhynchite sale in advance guaranteed quality until the end of Tybi (January). Although large-scale microbial destruction (due, e.g., to acetobacter or lactobacillus) is rare in the vintages of contemporary wine hobbyists (to say nothing of the modern "industrial" vineyard), the *ampelourgos* was significantly handicapped by his equipment and his limited knowledge of what would now be classified as chemistry and microbiology. In the end, however, I would prefer to suggest that ἀποκλείω does not have a negative sense, i.e., that it only references the segregation of two products and that the *oxos* in question was desired from the start; vinegar was hardly a worthless commodity (cf. Grant 1997, 288). This may be confirmed by 2480, which doubtless postdates 2044 by two indictions. In lines 124–25 of the former text, I read (on a photograph), ὡς λοιπ(ὰ) οἴν(ου) δι(πλᾶ) λ[δ]d´ [= surplus for present, fourteenth indiction] *vac.* ἐξ ὧν ἀπὸ πλεί(ονος) [= surplus *expenditure*] λόγ(ων) ιγ καὶ ιβ´´ οἴν(ου) [ο < ι] δι(πλᾶ) σιβγ´´ *vac.* ὡς πλεί(ονα) [again, a deficit] οἴν(ου) δι(πλᾶ) ϱοηιβ̣´ ἐνεχθ(έντα) | εἰς τὸν λόγ(ον) ιε ἰνδ(ικτίονος). π̣λ̣ή̣ϱης. If there had been a significant production failure in the twelfth indiction, one might reasonably expect more substantial arrears. (Incidentally, in 2480.123, read ἀνθ᾽ (ὧν) for ἀνθ(ελομένης).)

21. For this issue, see p. 16 and references (of which Palme 1998b is the most germane to the matter at hand).

22. Cf. Palme 1998b, 295 and n. 17.

23. Note, further, the discussion in *P.Thomas*, pp. 252–53.

24. The return to a more substantial *hypodochē* (ca. 30,000 *knid.*), which is attested under the Sasanians (*P.Oxy.* LVIII 3960), does not contradict this hypothesis, because in the end, Apion III received Apion II's Oxyrhynchite property; see pp. 16–17 above.

25. I owe this observation to Nick Gonis. With a date of 580/81 for 2480, the Brumalia distributions of wine hitherto connected with Apion II (2480.37–40; cf. Perpillou-Thomas 1993b) would be those of Apion III. This is not problematic, since both names begin with *alpha* and would have been celebrated on Hathyr 28. Indeed, τοῦ δεσπότου ἡμῶν τοῦ ὑπερφυεστάτου Ἀπίωνος, "of our master, the most extraordinary Apion," seems a more fitting reference to the young Apion III than to his distinguished grandfather (whose name, furthermore, probably would not have been written; cf., e.g., *PSI* III 191.1: τοῦ δεσπότου ἡμῶν τοῦ ὑπερφυεστάτου πατρικίου, "of our master, the most extraordinary patrician"). On Brumalia, see Crawford 1920 and Mazza 2005; one of the unpublished account fragments in Florence mentions Brumalia of the *despoina* (mistress).

The upshot of all this is that there was but one *oinocheiristēs* at a time, except, possibly, during the "interregnum" following Apion II's death. It is thus reasonable to believe that the accounts of these individuals usually represent the *global* traffic in wine on the estate, both inflows (rents, production, and purchases) and outflows (expenditures and sales).

Inflows

The structure of the annual accounts of the *oinocheiristai* is rather simple. The two "unabridged" examples (*P.Oxy.* XXVII 2480 and *PSI* VIII 953) primarily document distributions, which were set against the steward's *hypodochē*,[26] a quantity of wine (comparable in function to the total *lēmmata* in the pronoetic accounts) that the steward obtained for the most part from rents and any autourgic production. Only *P.Oxy.* LVIII 3960 attests these inflows, but in a summary fashion (ll. 4–18); presumably the text's aggregates were derived from receipts like those from the dossier of Philoxenos *oinocheiristēs.* On occasion, the *hypodochē* was augmented by purchases.[27] Explicit instances of such transactions appear in table 4.2.[28] *PSI* VIII 953.12–13 (4.2.4) may indicate that the steward purchased directly from the producers, that is, without a middleman; in any case, his direct and active involvement in the wine trade is most probable.[29] *P.Oxy.* LVIII 3954 might preserve an instance in which another manager, a *dioikētēs*, was involved in the acquisition of wine for the estate, but it is also possible that the transaction was personal. The same may be said of *P.Mich.* XV 743, where the buyer is an Apion notary. *P.Oxy.* XVI 2032.63 might be another case of a *notarios* buying on behalf of the *oikos.*[30] Given that Egypt's climate and soil are not suitable for the production of great wines,[31] the shipment of wine to Constantinople in 4.2.5 might seem noteworthy,[32] but the

Changing 2480's date to 580/81 may mean that the *hypatissa* (consular woman) referenced in line 19 of the text is Praeiecta and not Apion's widow. Praeiecta is, however, the only known *hypatissa* in the papyrological corpus; see also *P.Oxy.* XIX 2243a.86. Sarris (2006, 19) erroneously identifies her as "Apion III's wife." (His stemma on p. 23 is unreliable.)

26. Cf. *P.Oxy.* XXVII 2480.121–25, with the corrections in n. 20 above.

27. In 3960, purchased wine accounts for ca. 6 percent of the *hypodochē.*

28. I do not include the exaction in 3960.12.

29. Cf. p. 141 below.

30. See, further, p. 135 below.

31. Cf. the remarks in Rathbone 1991, 255–57.

32. Note also *PSI* VIII 956.17. Hardy (1931, 82–83) has suggested that the journey to Alexandria by the *kellarios* in *P.Oxy.* XVI 1904 (618) was for a similar purpose. (His implication is that the Apiones no longer maintained a residence in Constantinople at this time [see also his pp. 36–37], but cf., e.g., Gregory, *Reg. ep.* XIII 35.)

Apiones were probably not the only ones doing it.[33] Since the shipment is tagged ὑπὲρ κελλαρικῶν ("for stores"),[34] this particular wine was presumably not for sale. The only attestation of imported wine on the Oxyrhynchite estate is *P.Oxy.* XVI 1851, where Rhodian wine is mentioned; not surprisingly, it is being requested by a high-level estate official, the *antigeouchos.*

The prices extant in 4.2.1 and 4.2.2 are much too high (from the *oikos'* perspective) to have been sales in advance (i.e., cash loans with repayment in kind); "prices" in these agreements were typically in the 400–500 *sextarii* per *solidus* range in the Oxyrhynchite.[35] It is clear, however, that the estate used such contracts to procure wine,[36] possibly more often than is immediately apparent: a number of the Oxyrhynchite sales in advance involve buyer-creditors who are called *thaumasiōtatos* ("most admirable"),[37] a title generally borne by low-level managers like the *oinocheiristai* and *pronoētai.*[38]

Outflows, Part I: *Analōmata*

Although *P.Oxy.* LVIII 3960 dates from the Persian occupation and bears an incomplete list of *analōmata,* its summary categories nonetheless provide a useful structure for the analysis of our first "outflow" classification, that encompassing the Apiones' expenditures in wine. The other two well-preserved accounts, *P.Oxy.* XXVII 2480 and *PSI* VIII 953, present these expenditures in great detail, and the number and difficulty of their entries require that the

33. Still, the volume of Egyptian wine exports should not be overstated (see Wickham 2005, 713–14), and Van Minnen's assertion (2004, 438) that "Egyptian wine merchants indeed dominated the wine trade in late antiquity" is curious. The oft-cited Luzzatto 1996 (regarding *T.Varie* 3) actually concerns the movement of wine *within* Egypt (demonstrated conclusively by Federico Morelli in "Vino sul Nilo," *ZPE* 175 [2010]: 209–23).

34. Cf. Husson 1983, 147.

35. For typical advanced sale prices, see Kruit 1992a, 182–83. The lack of a loan element in 4.2.1 is noteworthy, since the *origo* of the sellers was an estate property, an environment in which we might expect greater dependency. The prices in 4.2.1 and 4.2.2 suggest that 8-*sextarius* containers (*knidia*) are used in this account; if not, the tariff would be too high for even a straight sale.

Rea, the editor of *P.Oxy.* LVIII 3960, is almost certainly correct in favoring a commuted rent (as opposed to a sale in advance) in 3960.36–38; note also p. 78 above. In 3960.36–38, 12 *solidi* correspond to 600 *knidia,* and this equation accords well with other estate data (12 *sol.*/ca. 3 *sol.*/*ar.* [cf. *apotakton chōriōn*] ≈ 4 *ar.* x 150 *d*/*ar.* [cf. the fixed-rate rents in Text I] ≈ 600 *d* [= *knidia;* cf. p. 192]).

36. Cf. Benaissa forthcoming.

37. See, e.g., *P.David* 10 (= *SB* X 10285); *P.Flor.* I 65; *P.Hamb.* III 221; *SB* V 8264, XIV 11617 (also has the private gold standard typical of the large estates).

38. For *thaumasiōtatos* with *oinocheiristai,* cf. *P.Oxy.* XVI 1951 and *SB* XX 14644 (which perhaps does not reference an *oinocheiristēs*). Neither of these pertains to the Apion estate, but there is no reason to think that the titulature that they attest is exceptional. For the title in reference to estate *pronoētai,* cf., e.g., *P.Oxy.* LVIII 3952.41.

TABLE 4.2. Purchases of Wine

Text (Date)	Quantity of Wine (Source)	Price	Notes
1. *P.Oxy.* LVIII 3960.13 (621)	450 *knidia* (παρὰ τῶν ἀπὸ Πηλέας [an *epoikion*])	25 *solidi* or 18 *knid./sol.*	
2. *P.Oxy.* LVIII 3960.14 (621)	864 *knid.* (παρὰ τῶν ἀπὸ Ἀπόλλωνος [a village])	36 *sol.* or 24 *knid./sol.*	
3. *P.Oxy.* LVIII 3960.15 (621)	800 *knid.* (παρὰ τῶν ἀπὸ Πουσεμπόϋς [a village])	?	—117 of the 800 returned (vinegar)
4. *PSI* VIII 953.12–13 (567–68)	20 ? *dipla,* [ἅπερ ἠγόρα(?)]σεν ὁ οἰνοχειριστὴς παρὰ τῶν ἀπὸ Βακιν (ed. pr. Βακι ̣, but cf. BL inv. 2219.12)	?	—purchased ὑπὲρ τοῦ μοναστηρίου Πέλα
5. *PSI* VIII 953.72–74 (567–68)	2,000 *dipla* + a 3% addition labeled ὑπὲρ ρρ( );[a] παρὰ τῶν ἀπὸ Ἰβιῶνος (a village)	?	—sent ἐν Κωνσταντίνου πόλει τῷ δεσπότῃ ἡμῶν τῷ κυρω [read κυρίῳ] ὑπὲρ κελλαρικῶν α ἰνδικτίονος

[a]The editor suggests *roga,* but I prefer *rhysis,* i.e., *hyper rhyseōn.*

investigation be painstaking. Close analysis seems all the more essential in light of the misunderstandings (and omissions) of previous commentators.

Pious and Charitable: λόγῳ προσφορᾶς, λόγῳ εὐσεβείας

The expenditures section of *P.Oxy.* LVIII 3960 commences (ll. 20–22) with an entry for wine distributed "to the holy churches and hostels and *martyria* of the city and in the country on account of offering according to a schedule which is in the secretariat."[39] The sum recorded, including vinegar,[40] is a whopping 8,458 *knidia*—just over 25 percent of the steward's *hypodochē* for the indiction. Attestations of *prosphorai* (offerings) in wine from elsewhere in the dossier are listed in table 4.3. The small quantities are striking: The largest of

39. εἰς τὰς ἁγίας ἐκκλησίας καὶ ξενοδοχεῖα καὶ μαρτύρια τῆς πόλεως καὶ ἐν ἄγροις λόγῳ προσφορᾶς ἀκολούθως βρεβίῳ ὄντι ἐν τῷ χαρτουλαρίῳ. Distributions to religious institutions also begin the *analōmata* in the accounts of the *pronoētai.*

40. Rea, the editor, mistranslates *oxos* as "new wine."

TABLE 4.3. *Prosphorai*

Text (Date)	Agent	Recipient	Amount	Notes
1. *P.Laur.* IV 178 (557 or 572)	Serenos *oinocheiristēs*, through ?[a]	Hagios Theonas	10 *dipla*	
2. *P.Oxy.* X 1322 (413)	Petros *boēthos*, through Lamason *ampelourgos*	Menas priest	94 *dipla*	—Included here on the chance that this Lamason also appears in XVI 2043[b]
3. *PSI* VIII 953.30 (567/68)	Serenos *oinocheiristēs*	Hagios ?	120 *dipla*	—"By order of our master and lord"
4. *PSI* VIII 953.31 (567/68)	Serenos *oinocheiristēs*	Hagia Euphemia	52 *dipla*	—"By order of our master and lord" —Recipient located in the city of Oxyrhynchus[c]
5. *PSI* VIII 953.82–83 (567/68)	Serenos *oinocheiristēs*	(St.) John the Evangelist	at least 72 *dipla*	—"By order of our master and lord"
6. *PSI* VIII 955.19[d]	? *oinocheiristēs*	?	?	—opaque; a nilometer also seems to appear in entry

[a]According to the editor, the order is addressed τῇ ἀδελφ(ότητι) Ἰσιδόῤῳ [read Ισιδώρῳ] Πατωσαρ(έτ). In the DDBDP, ἀδελφ(όντι) appears. I would prefer ἀδελφ(ῇ), with either Isidora in the dative or the genitive of Isidoros following. The plate in the volume is of no help to me.

[b]Cf. appendix B, s. ΠΕΤΡΩΝΙΟΥ.

[c]Cf. *P.Oxy.* VII 1038.

[d]Date of *PSI* VIII 955: Probably 570/71, cf. p. 91, n. 4.

them, the 120 *dipla* in 4.3.3, would have had to have been made seventy times to equal the amount distributed in 3960. That the number of attestations for *prosphorai* is considerably less than this is probably significant; in other words, it is likely that the figure in 3960 is atypical. The *prosphorai* in 4.3.3–5 are described as "by order of our master and lord." This suggests that *prosphorai* need not have necessarily been permanent (or customary) distributions, that they also could have been occasional gifts—potentially *kata to ethos*, but not inevitably so. It follows, then, that the schedule (*brevium*) that governed the *prosphorai* in 3960 was not a fixed document (as, e.g., the *apaitēsimon*) but an evolving and variable text. No doubt relevant for the quantity of wine in 3960 is the fact that *prosphora* frequently denotes an offering for masses for the dead.[41] Egyptian custom specified that memorial masses be celebrated on the

41. For the meanings of *prosphora*, see *P.Oxy.* LXVII 4620, 2–3n. and refs.

third, seventh, and thirtieth day after decease, as well as on its anniversary for many years thereafter, sometimes permanently.[42] Surely the estate's *prosphorai* were typically not as large as they are in 3960; they reflect the recent (and probably premature) death of Apion III.[43]

Of course, not all *prosphorai* were mass offerings. In its broadest sense, the term means "any kind of pious gift,"[44] and it is clear, for example, that *nosokomeia* (hospitals) received them,[45] and that *prosphora* might also refer to the salary of a cleric.[46] The absence of monasteries in the heading in 3960 is notable; it is echoed by the attestations in table 4.3, none of which explicitly involves a monastery. Though *prosphorai* were certainly given to monasteries in the region,[47] the evidence from the Apion dossier would seem to support Wipszycka's criticism of Steinwenter's assertion that monasteries were the recipients of choice for offerings for the dead.[48]

The entry for *prosphorai* in 3960 is followed by one (l. 23) reading "to the sanctified monasteries and the widows and invalids on account of piety."[49] These distributions amount to 576 *knidia,* about 2 percent of the steward's *hypodochē.* Table 4.4 lists other Apion distributions of wine that are explicitly *pietatis causa.*[50] The distributions in this table are consistent with the heading in 3960; they too are limited to widows, the disabled, and monasteries. The role of monasteries in the care of those on society's margins is a link between the recipients.[51] Table 4.4.1 and table 4.4.3–5 are described as "customary," but the other five distributions lack this designation, and no pattern regarding frequency is apparent. The entries *logō eusebeias* in *P.Oxy.* 2480 (4.4.1–7) amount to 393 *dipla* (349 vinegar + 44 wine), or just over 2 percent of the steward's *hypodochē;* that this percentage is the same as that in 3960 may indicate some

---

42. Wipszycka 1972, 69–70.

43. Note also *P.Oxy.* LXVII 4620, where the possibility of such an offering for "a deceased lady of the Apion house" is mentioned. *Prosphorai* for "our mistress [*despoina*]" appear in one of the unpublished wine account fragments in Florence.

44. Wipszycka 1972, 65.

45. Cf. *P.Oxy.* LXI 4131 (*sitos*). In 3960, they may well have been covered by the term *xenodocheion;* cf. Wipszycka 1972, 115. For the use of wine in monastic health care, cf. Crislip 2005, 28–29.

46. See *P.Oxy.* LVIII 3936.18–19, where a priest's salary (*opsōnion*) is described as ἐξ ἔθους διδουένου ὑπὲρ τῆς ἁγίας προσφορᾶς, "given by custom for the holy offering."

47. Cf., e.g., *PSI* I 89 (*sitos*) and VII 786 (money).

48. Wipszycka 1972, 75–77.

49. εἰς τὰ εὐαγῆ μοναστήρια καὶ ταῖς χήραις καὶ ἀδυναμοῦσι λόγῳ εὐσεβείας.

50. Unpublished witnesses—such as the two that appear in a Florentine fragment to be published by Mazza and myself—have not been included.

51. Cf. Talbot 1990, 121, and Richter 2005, a masterful analysis of the Coptic child donation contracts.

TABLE 4.4. Distributions *Pietatis Causa*

Text (Date)	Agent	Recipient(s)	Amount	Notes
1. *P.Oxy.* XXVII 2480.5 (580/81)[a]	Phib *oinocheiristēs*	Petros, *apo symmachōn,* blind	36 *dipla*	—labeled "customary" —a certain Georgios receives 280 *folles* ἀντὶ Πέτρου πολλοβλέπτου λόγῳ εὐσεβείας ἐξ ἔθους in *P.Oxy.* XVI 1921 (621)
2. *P.Oxy.* XXVII 2480.12 (580/81)	Phib *oinocheiristēs*	Heraïs, wife of Anoup	8 *dipla*	—Heraïs' link to *oikos,* if any, unknown
3. *P.Oxy.* XXVII 2480.31–32 (580/81)	Phib *oinocheiristēs*	*koinobion* of Abbas Iere-mias in the Thebaid	200 *dipla* vinegar, of which 124 were commuted to coin	—labeled "customary"
4. *P.Oxy.* XXVII 2480.44 (580/81)	Phib *oinocheiristēs*	*xenodocheion* of Abbas Arion	80 *dipla* vinegar	—labeled "customary"
5. *P.Oxy.* XXVII 2480.46 (580/81)	Phib *oinocheiristēs*	*koinobion* of Abbas Hermes	13 *dipla* vinegar	—labeled "customary" —*PSI* I 89 (605) = customary distribu-tion of wheat to same place λόγῳ προσφορᾶς
6. *P.Oxy.* XXVII 2480.119 (580/81)	Phib *oinocheiristēs*	*monastērion tou Orous*	6 *dipla* vinegar	
7. *P.Oxy.* XXVII 2480.120 (580/81)	Phib *oinocheiristēs*	*koinobion* of Abbas Pamoun ἐν τῷ Ὄρι [read Ὄρει] Πάεε	50 *dipla* vinegar	
8. *PSI* VIII 953.61 (567/68)	Serenos *oinocheiristēs*	Sibylla?, a widow	4 *dipla*	
9. *PSI* VIII 953.63 (567/68)	Serenos *oinocheiristēs*	?, presumably ἀπὸ] συμμάχων	36 *dipla*	—Petros here? cf. 1

[a]For this account's date, see pp. 95–97.

TABLE 4.5. Miscellaneous Pious and Charitable Distributions

Text (Date)	Agent	Recipient(s)	Amount	Notes
1. *P.Oxy.* XVI 2044.18 (prob. 579)[a]	Phib *oinocheiristēs*	the *kellion* of the *xenodocheion*	1,000 *dipla* vinegar	—Not clear if *kellion* belongs to *xeno-docheion* or is simply named after it (latter seems more likely)
2. *P.Oxy.* XXVII 2480.84 (580/81)[b]	Phib *oinocheiristēs*	those in the *domus gloriosa*'s *phylakē* on Easter	2 *dipla*	
3. *P.Oxy.* LXVII 4622 (V/VI)	Biktor *oinocheiristēs*	widows of the Martyrium of St. John	1 *diploun*	—connection to Apiones not certain
4. *PSI* VIII 953.6 (567/68)	Serenos *oinocheiristēs*	the Mega Oros (monastery)[c]	100 *dipla* vinegar	
5. *PSI* VIII 953.8 (567/68)	Serenos *oinocheiristēs*	*ekklēsia* Papsau	20 *dipla*	
6. *PSI* VIII 953.9 (567/68)	Serenos *oinocheiristēs*	*koinobion* Abbas Sarmates	100 *dipla* vinegar	
7. *PSI* VIII 953.9 + *BL* VII 238 (567/68)	Serenos *oinocheiristēs*	*xenodocheion* Amma Bes( )	?	
8. *PSI* VIII 953.10 (567/68)	Serenos *oinocheiristēs*	Hagios Serenos	at least 20 *dipla*	—a *martyrion*, cf. Papaconstantinou 2001, 274–75
9. *PSI* VIII 953.11 (567/68)	Serenos *oinocheiristēs*	*ekklēsia* of Abbas Hierakion	16 *dipla*	
10. *PSI* VIII 953.11 (567/68)	Serenos *oinocheiristēs*	Theodosia	at least 14 *dipla*	—a widow?
11. *PSI* VIII 953.12–13 (567/68)	Serenos *oinocheiristēs*	*monastērion* Pela	360 *dipla* + at least 60 *dipla* vinegar	—a portion of these (at the least) were purchased
12. *PSI* VIII 953.37 (567/68)	Serenos *oinocheiristēs*	those imprisoned in the *phylakē* of the *endoxos oikos* on Easter	1 *diploun* vinegar	

TABLE 4.5—*(Continued)*

Text (Date)	Agent	Recipient(s)	Amount	Notes
13. *PSI* VIII 953.54–55 (567/68)	Serenos *oinocheiristēs*	for 139 in the *phylakē* of the *endoxos oikos* ἐν τῇ ἁγίᾳ ἡμέρᾳ τοῦ ἀρχαγγέλου Μιχαήλ	20 *dipla*	
14. *PSI* VIII 953.56–57 (567/68)	Serenos *oinocheiristēs*	those in the *phylakē* on Easter, three days	40 *dipla* + 40 more *dipla* used for the purchase of meat and *opsarion*	
15. *PSI* VIII 953.59–60 (567/68)	Serenos *oinocheiristēs*	those in the *desmōtērion* ἐν τῇ ἑορτῇ τοῦ Τῦβι μηνός	30 *dipla*	

[a] For this text's date, see pp. 96–97.
[b] For this account's date, see pp. 95–97.
[c] For the reading, see *BL* XI 248.

stability in the estate's distributions *pietatis causa.*[52] Also noteworthy is the fact that all of the monasteries receive vinegar instead of wine, which is suggestive of a later monastic dietary triad consisting of vinegar (ϩⲙⲝ), salt, and dry bread.[53] Vinegar curiously does not appear in the *pietatis causa* entry in 3960 (but it does, strangely enough, in the preceding entry for *prosphorai,* though none of the other Apion *prosphorai* are in vinegar).

Finally, there are references that would seem to belong under the rubrics presently at issue but that are not labeled explicitly. These are listed in table 4.5 and come predominantly from *PSI* VIII 953. The absence of categorization could either be a quirk of the steward or represent an earlier form of bookkeeping practice (since 953 is the earliest account). Monastic recipients of vinegar (4.5.4, 6, and 11) are very probably *pietatis causa;* the fact that 953 lacks explicit monastic entries *logō eusebeias* supports this conclusion. In contrast,

52. *P.Oxy.* 2480 is admittedly incomplete, but only the distribution entries for Mesore seem to be missing. The steward's fiscal year apparently ran from Mesore through Epeiph; i.e., it (sensibly) started with the vintage, not in Thoth or Pauni.

53. See Nagel 1983. In the period under consideration here, wine was, as a rule, limited to monks who were sick; cf. Crislip 2005, 74–75.

4.5.5, 4.5.8, and 4.5.9, all concern wine given to churches.[54] These could be *prosphorai*, but if so, it is curious that they are not labeled as such, since that term appears elsewhere in the account; perhaps the steward did not use it for *prosphorai sensu lato*. Yet the permanence of the gift might also factor in the distinction: "By the order of . . ." does not appear in 4.5.5, 4.5.8, and 4.5.9 but is used by this steward for his distributions that are tagged as *prosphorai*. These possibilities are not mutually exclusive, of course. Table 4.5.7 is an uncertain case, all the more so because it is not clear whether wine or vinegar was distributed; *xenodocheia* appear in 3960 as recipients of *prosphorai*, but table 4.4.4 is a customary distribution of vinegar to a *xenodocheion* that is tagged "*eusebeia*." Theodosia (4.5.10) could be a widow and therefore a recipient *pietatis causa* (as the widows in 4.5.3 certainly are), but if so, it is surprising that she is not identified as such (cf. table 4.4.8); although the overwhelming majority of entries in this column of 953 fall under the heading of pious and charitable, this one need not. Presumably the distributions to prisoners on holy days and festivals (4.5.2, 12–15) were considered acts of charity.[55] Table 4.5.1 is ambiguous; the *kellion* of the *xenodocheion* may well have been named for its proximity to said institution.

Ὑπὲρ συνηθειῶν

Line 24 of *P.Oxy.* LVIII 3960 reads "to the *taxeōtai* for *synētheiai*, by custom also for the 9th indiction,"[56] 336 *knidia*. The amount is small, almost exactly 1 percent of the steward's *hypodochē*. The *taxeōtai* in question were no doubt members of the bureau (*officium;* Greek *taxis*) of the governor (*praeses*) of the province of Arcadia;[57] the conquering Sasanians had lopped off the head but left the body intact.[58] The *synētheiai* (here in a narrow sense, "perquisites," Latin *sportulae*) were, by this period, removed from bribes and illicit gratuities. They had lost their personal aspect and had been incorporated into the fiscal system; they were subject to imperial legislation. This public character is quite clear in the fiscal accounts of Aphrodite, in which they are classified among the *dēmosia* (taxes).[59] At Oxyrhynchus, their fiscal nature is suggested by the ex-

54. Table 4.5.8 is more accurately a *martyrion;* cf. Papaconstantinou 2001, 274–75.
55. Cf. Torallas Tovar 2006, 109–10.
56. τοῖς ταξεώταις ὑπὲρ συνηθειῶν ἐξ ἔθους καὶ ἐπὶ τῆς θ ἰνδικτίονος.
57. For an overview of the administration during this period, see Palme 2007.
58. Palme (2007, 265) notes, "One has the impression that the Sassanids conquered Egypt by surprise but afterwards made no attempt to replace the existing state structures."
59. Cf. Zuckerman 2004, 115.

istence of receipts for their payment by agents (*epimelētai*) of the house of Theon.[60]

The house of Theon paid its *synētheiai* in coin; in fact, save for the entry in 3960 with which this section began, there are no explicit attestations of officials' *synētheiai* in wine from the Oxyrhynchite or, to the best of my knowledge, elsewhere. The singularity of 3960 conceivably could be tied to the Sasanian occupation,[61] but there exist unspecified distributions of wine that might be *synētheiai*. In *P.Oxy.* XVI 2047 (V/VI),[62] for example, the well-known counts Phoibammon and Samouelios order that two *singulares* be given an *omphakēra* of *propoma* (an aperitif), ten *knidia* of wine, six hens (*ornea*), sixteen pounds of meat, and thirty white ("clean," *kathara*) loaves;[63] *singulares* were part of the governor's bureau and are among the recipients of *synētheiai* in *P.Cair.Masp.* I 67054.[64] From the Apion dossier itself, there is *PSI* VIII 953.7, which records a distribution of twenty-four *dipla* to an *adiutor* (Greek *boēthos*, "bureau assistant, [usually, but not necessarily] with a more military nuance");[65] here it may be noted that an *adiutor delegationis* receives a cash payment for *synētheia* on behalf of his bureau (*scrinium*) in one of the Theon texts.[66] Line 10 of 953 has a payment of wine to a *subadiuva*, a "deputy assistant" or *hypoboēthos*,[67] while the fragmentary 955.38 records an allotment of wine to a pagarch.[68] The latter might not be a *synētheia* payment, however, for

60. See *P.Oxy.* XVI 1999, 2009. For the *oikos* of Theon, see Gascou 1985, 41–46, and Gonis 2009. When the receipts were drafted, it was a purely fiscal entity.

61. One might posit that payment in kind seemed preferable to the *oikos* because gold was increasing in value under the invaders; yet 3960.13–15 show a willingness to make purchases that were at least valued in gold. Maresch 1994, 72–74, discusses the tariff of the *solidus* during the period in question.

62. HGv dates to V.

63. For the *comites* Phoibammon and Samouelios, see *P.Oxy.* LXVIII 4697. Concerning the *omphakēra*, see appendix C, s.v.

64. For *singulares*, see *P.Oxy.* LVIII 3932, 3n.; *P.Thomas* 27, 1n. Note also *P.Mert.* II 95 + *BL* VII 105.

65. The quotation is from *P.Sorb.* II 69, p. 250 (110.D10). The *adiutor* in question probably also appears in *PSI* VIII 956.29.

66. *P.Oxy.* XVI 2009. In line 2, the editors resolve σκϱ( ) as σκϱ(ίβα), "clerk," stating that it "seems more likely than σκϱ(ινιαϱίου) . . . in connexion with the *embole.*" σκϱ(ινίου) is probably correct; cf. *P.Cair.Masp.* I 67057/2.11 + *BL* IV 13, where τοῦ ἀν[νων]ιακ(οῦ) σκϱ(ινίου) appears. A *boēthos* of the same appears in the next line of this text. For the *delegatio divina* as a recipient of *synētheiai*, see *P.Cair.Masp.* I 67054/1.13.

67. Cf. *P.Sorb.* II 69, p. 253 (122.B1). The same *subadiuva* appears a decade later in *P.Oxy.* VII 1042 (578). This connection was already proposed by Robinson in *CPR* XIV 39, 10n., though her statement that "a praesidial official was at one time on a private pay roll" suggests an overly rigid view of public and private in late antique Egypt.

68.]ạ here is clearly the end of μετὰ, not Μηνᾶ (contra, e.g., Mazza 1995, 241). See the next note.

953.51 concerns distributions to soldiers (precisely, *buccellarii*) who are traveling with an individual bearing a title typical for a pagarch;[69] in other words, the wine in 955 could have been for rations (cf. the following section on the *annona militaris*).

Yet not all of the *synētheiai* encountered in the papyri are payments to officials. They appear, for example, as perquisites for the landlord in leases,[70] and a late Oxyrhynchite work contract for an overseer specifies that he will receive "all the *synētheiai* that said overseer is accustomed to receive annually from the agriculturalists, be they in grain or in wine or in other forms."[71] In *P.Oxy.* 2480, wine distributions to carpenters working on the irrigation machinery of Oxyrhynchus (*tektones mēchanourgoi*) are classified as *synētheiai* (l. 45); eight *dipla* are given, and the allotment is described as customary.[72] Perpillou-Thomas links this distribution to the *heortika* (festival "extras") that appear in service contracts.[73]

Another kind of festal distribution is fittingly mentioned in this section. These are the *kalandika,* or New Year's gifts,[74] which are paired with *synētheiai* in *Edict* XIII.3; this chapter of the edict specifically mentions the *synētheiai* and *kalandika* that the Augustal duke of Egypt (*dux et Augustalis Aegypti*) was to receive *through* the *vindex* of Alexandria.[75] One also encounters officials receiving (or complaining that they have not received) *kalandika* in the papyri.[76] *Kalandika* in wine do appear in *P.Oxy.* 2480.41–43, though they are not for

69. The text only has]αρρ( ), but the restoration βουκκελλ]αρρ( ) is without doubt. Concerning their leader's titulature, see Gascou 1972, 69 and especially n. 2 (an argument needlessly abandoned wholesale in Gascou 2008). The same titulature (and little else) reappears in *PSI* VIII 956.20 and thrice in an unpublished fragment in the Laurenziana (to be edited by Mazza and myself). In 956.20 (and once in the ineditum), it is attached to the name Menas. This may well be Flavia Anastasia's father (for Anastasia, see *P.Oxy.* LXIX 4756–58 and refs.).

70. See Herrmann 1958, 119–20; *P.Thomas,* pp. 51–68.

71. *P.Oxy.* XIX 2239.18–20: πάσας τὰς συνηθείας ἃς εἴωθεν λαβεῖν ὁ αὐτὸς ἐπικείμενος κατὰ τὸ ἔθος παρὰ τῶν γεωργῶν εἴτε ἐν σίτῳ καὶ ἐν οἴνῳ καὶ ἐν ἄλλοις εἴδεσιν.

72. Perhaps cf. the distributions to bath attendants (*perichytai*) in two texts from the Fayum (*SB* I 4905 and *Stud.Pal.* VIII 1110), if the baths in question were public.

73. Perpillou-Thomas 1993a, 246ff., especially 249 and n. 29. For bonus payments in labor contracts, cf. *P.Heid.* V, pp. 158–59.

74. Cf. the discussion in *P.Thomas* 27.

75. I emphasize "through" because Johnson and West (1949, e.g., 249) are misleading, implying that the *vindex* (an official supervising tax collection) was the original source. The text reads καὶ ὑπὲρ συνηθειῶν καὶ καλανδικῶν παρὰ βίνδικος ἐκ τῶν ὑποτεταγμένων τῷδε τῷ τύπῳ τόπων τε καὶ αἰτίων (I prefer this to αἰτιῶν; cf. Thurman 1964, 141 n. 300), "and for *synētheiai* and *kalandika* through the *vindex* from the places and responsible parties appended to this edict."

76. E.g., *P.Oxy.* XVI 1869 (*scriniarius*) and 1875 (very probably an official; he is leasing house property, and one frequently encounters such individuals in that context). Note also the *kalandarika* in *P.Thomas* 27.

officials but for mimes and acrobats, probably some public holiday entertainment. Perpillou-Thomas has no doubts, noting, "These artists are clearly employed for an official ceremony . . . and not for private festivities; they can therefore be regarded as public contractors."[77] Yet *kalandika* clearly can appear in the private sphere. How else might one explain *PSI* XIV 1428.12–13, in which a master is informed about "the *kalandika* for the sake of your household" (τὰ καλανδικὰ εἰς λόγον τῶν οἰκετῶν ὑμῶν)? Perpillou-Thomas herself is uncomfortable with her suggestion that this letter comes from an administrative context.[78]

Officials might also receive payments *logō philotimias*,[79] but such payments were not limited to them, nor were they always in wine. The two extant wine distributions of this type that involve officials come from *P.Oxy.* 2480. The first concerns the governor (*archōn*) Thomas, who receives 100 *dipla* on his entry into office (or possibly on the occasion of a visit to the estate);[80] in the second, the *prōtodēmotai* of Oxyrhynchus (who were liturgists) are given 20 *dipla* on the birthday of an Apion *femina consularis*.[81]

Because it likewise concerns a payment to an official, the entry in *P.Oxy.* LVIII 3960 following that for *synētheiai* (l. 25) should be addressed here. It reads, "to the *embolatōr*, 100 *knidia*, and the *meizoteros*, 150 *knidia*, with freight charges for tenders, 42 *knidia*, and for others, 260 *knidia*, total 322 [*sic;* 552 should appear] *knidia*";[82] the correct sum is a little over 1 percent of the steward's *hypodochē*. The presence of a charge for freight (*naulon*), often a surtax on the *embolē*, suggests that this entry concerns the transport of annonary grain (and not of wine).[83] The "others" are also

77. Perpillou-Thomas 1993a, 261, following Gascou 1976b, 195 n. 2.

78. Perpillou-Thomas 1993a, 263. There is no reason to see *kalandika* in *P.Oxy.* XVI 1950, a receipt for a distribution of wine "on account of the festal gifts of Tybi" (λόγῳ ἑορτικῶν Τῦβι, i.e., December–January) from the church to a *boēthos exactorias.* The text dates to 24 January, and Epiphany or another feast is probably behind the distribution; cf. Perpillou-Thomas 1993a, 255. (Perpillou-Thomas' statement, however, that Tybi 1 "is not the date of a great Christian feast but calls to mind the importance of the νεομηνίαι [first days of months] in the pagan religious mentality" seems somewhat hasty; cf. Papaconstantinou 1996, 146.)

79. For the meanings of *philotimia,* see Morelli 1998, 220–21. For *P.Oxy.* 2480, Morelli suggests rendering it as "largesse" (contra MacCoull 1990).

80. *P.Oxy.* XXVII 2480.62; the wine is given to Thomas ἐν τῇ εἰσόδῳ αὐτοῦ. He is not to be confused with the homonymous *praeses Thebaidis* (for whom see Fournet 1998, 79–80).

81. *P.Oxy.* XXVII 2480.18–19. As indicated in n. 25 above, I would argue that this *consularis* is Praeiecta. For the *prōtodēmotai,* see Gascou 1976b, 201–12, and *P.Oxy.* LIX 3987, 7n. (where Gascou's extensive discussion is not mentioned). At 3987.11–12, the position is explicitly called a *leitourgia.*

82. τῷ ἐμβολάτορι 'κνίδια ρ' καὶ τῷ μειζοτέρῳ 'κνίδια ρν' σὺν ναύλοις πακτωναρίων 'κνίδια μβ' καὶ ἄλλοις 'κνίδια σξ' οἴνου κνίδια τκβ.

83. Could the use of smaller boats (*paktōnaria*) be connected to the diminution (and geographic shift) of annonary demands under the Sasanians?

explainable,[84] though it is curious that both charges are in kind. The *meizoteros* might be a commander whose soldiers (*buccellarii*) are protecting the shipment;[85] otherwise, he is likely to be a village headman.[86] Though the general responsibility of the *embolatōr* might seem obvious from his title, he is an obscure figure—but not as obscure as some have suggested.[87] Within the Apiones' sphere at least, he was clearly *not* the collector of the *annona civica;* the *pronoētai* made their deliveries directly to the boats.[88] But the *embolatōr* does purchase surplus grain from the *pronoētai* in *P.Oxy.* XVI 1911.208–9 and 1914.4–7. In the latter account, it is also recorded that he delivered this grain to a boatman (*nautēs*), no doubt for inclusion with the *embolē;* perhaps, then, he did not accompany his purchases downriver. *P.Oxy.* I 126.14–16 would seem to indicate that he was an agent of the praetorian prefect; there an *embolatōr ētoi arkarios* is due to receive "twenty-two and a half carats of pure gold, the equivalent of twenty-four carats on the public standard" (χρυσοῦ κεράτια εἴκοσι δύο ἥμισυ ὀβρυζιακὰ εἰς δημοσίῳ κεράτια εἴκοσι τέσσαρα) for *arkarika,* revenues due the praetorian prefect.[89] Of course, such a position would not preclude an oversight role with the *embolē,* since collection and transport of the *annona civica* were under the supervision of the prefect.[90] The money payment for *arkarika* is perhaps echoed in *P.Oxy.* XVI 1908.26–28, where the *embolatōr* receives over 105 *solidi,* the bulk of which are earmarked for the second *katabolē* (tax installment). In the following line of this text, he is paid 9.25 *solidi* on the public standard for *synētheiai,* and this disbursement is paralleled by *P.Cair.Masp.* I 67054/1.7 and 67057/2.2, as well as in *P.Oxy.* XVI 1999, where

84. For these "others," cf. *P.Cair.Masp.* I 67057/3.13 with Gascou 1989, 298, 306–7, and 313. Individuals are certainly not meant.

85. Cf. *PSI* VIII 953.68–69, 78 (referring to shipments of gold taxes). For *meizoteroi* leading *buccellarii,* see p. 119 below. "Collaboration" with the occupying Persians may also occur in *P.Oxy.* LVIII 3960.37–38; cf. n. 92. But it might be less appropriate to make this charge against *buccellarii* than against "regulars."

86. *Meizoteros* is polysemous. Village headman: See *P.Oxy.* XVI 2018, and note 2021 (an account of the *embolē* of the village Takona in which a *meizoteros* receives 50 *artabai* of grain).

87. Cf. *P.Oxy.* LVIII 3960, 25n. ("his functions are not at all clear") and *BGU* XVII 2729, 5n. ("his position within the administration cannot be determined exactly"). Johnson and West (1949, 327) curiously assert that the *embolatōr* was replaced by the *osprigitēs* at a later date; concerning this, cf. *P.Oxy.* LVI 3873, 8n. Sarris (2006, 77) writes of "the *embolator* (ἐμβολάτωρ), or *demosios nautes* (δημόσιος ναύτης), who took charge of the Apion family's share of the imperial *embole* (ἐμβολή), or corn supply." His basis for this definition (and for putting the *embolator* in his fig. 5, showing "the administrative structure of the Apion household") completely escapes me.

88. Cf., e.g., *P.Oxy.* LV 3804.149–50.

89. For *arkarika,* see Rouillard 1928, 93–95, and Johnson and West 1949, 302–3. *P.Oxy.* XVI 2020 is an account τῶν ἀρκαρικῶν τίτλων (fiscal charges owed to the praetorian prefecture). For the nuances of *ētoi,* see P. J. Sijpesteijn 1992.

90. See, e.g., Jones 1964, 698.

two *embolatores* receive 7 *solidi* less 14 *keratia* from the *oikos* of Theon *hyper synētheiōn embolōn*. The distribution of wine in 3960 might be for such *synētheiai;* at the tariffs in that account, the 100 *knidia* received by the *embolatōr* would be the equivalent of between 2 and 5.56 *solidi.*

Annona Militaris

P.Oxy. LVIII 3960 has several entries mentioning Persians, and the vast majority of the wine in them was no doubt consumed by the occupying army. Line 26 records 7,281 *knidia* "distributed to the Persians according to a register in the secretariat,"[91] just under 22 percent of the *oinocheiristēs' hypodochē.* Presumably 498 of these *knidia* were those that had been exacted (*apaitēthenta*) from the inhabitants of Episemou "on account of the Persians" (l. 12). A distribution "to the tribune for the rations of the Persians" (τῷ τϱιβούνῳ εἰς τὴν ῥόγαν τῶν Πεϱσῶν) appears in lines 37–38,[92] but it is not clear if it was actually made in wine or adaerated (commuted to cash).[93]

Wine, bread, meat, and oil were the standard components of the rations allowance of the late antique soldier.[94] Though the Oxyrhynchite was hardly as militarized as some other regions of the East, payments of/for wine and other rations are prominent in the papyrological documentation stemming from the district. *Buccellarii* are the units most often encountered. These troops "are something of a *cause célèbre* in late antique history. The issue is whether they were soldiers, representatives of the state, or armed guards, representatives of the great estates."[95] One's stance is inevitably dependent on one's view of the relationship between the *oikoi* and the state. Those who discern centrifugal forces, even "feudalism," in the sources will incline toward private armies, toward interpreting these troops as an abuse of power that the emperors vainly attempted to proscribe.[96] Those who accept the fiscal model, in turn, will argue that *buccellarii* were imperial forces, that the estate personnel who supplied and directed them did so in their capacity as authorized

---

91. ῥογευθέντα τοῖς Πέϱσαις ἀκολούθως καταλόγῳ ὄντι ἐν τῷ χαϱτουλαϱίῳ. For *rhogeuō/rhoga*, an almost certain indicator of soldiers, see Kramer 1992.

92. For possible collaboration here, cf. *P.Oxy.* LVIII 3960, 37–38n.

93. The tribune received either 7.333 *solidi* or its equivalent in wine (366+ *knid.*). The latter seems more likely.

94. Mitthof 2001 is essential for military logistics in late antique Egypt but touches only briefly on the texts discussed in this section.

95. The words of Alston 2002, 413.

96. Hardy 1931, 60–67, especially 65–67, is a locus classicus; more recently, see Sarris 2006, 162–75.

agents of state power.[97] Though the present section is influenced by the latter outlook,[98] the estate account data themselves are often equivocal.[99]

The numerous military distributions in the Apion accounts may be more manageably assessed by text. Table 4.6 contains the entries from the steward's account *P.Oxy.* 2480. If *P.Oxy.* LVIII 3960 is a reliable guide, these distributions are discrepant; they amount to a mere 2.4 percent of total expenditures. Since 2480's section of dated distributions is nearly complete,[100] it is probable that the *oikos*' local contingent of *buccellarii* (a much greater expense, presumably) was addressed in the account's (now fragmentary) household distribution list.[101] It may also be that the presence of Sasanian troops necessitated greater expenditures in 3960. The majority of the entries in the dated portion of 2480 are for *buccellarii* and *neōteroi* ("recruits," the newly enrolled),[102] terms that apparently can function as synonyms here (cf. 4.6.1–4, 28–29). The Kromidiotai (4.6.10–18)—"onion-eaters" to the "doughboys" (i.e., the *buccellarii,* literally, "biscuit eaters")?—are somewhat more obscure.[103] Their name may in fact derive from their place of residence.[104] They may have been mounted,[105] and their lower rations (see next paragraph) may indicate that they were some sort of auxiliary force. For the most part, the *buccellarii* are said to have come from Kynopolis, Herakleopolis, and Koma (farther north in the Herakleopolite nome). Table 4.6.19–22 document missions of *buccellarii* from Oxyrhynchus; the relative scarcity and nature of these operations—all are away from the city—provide further support for the argument that this contingent was included in the household list. Unit size is small; we are very far indeed from the huge assemblages of *buccellarii* in Procopius. The highest concentration of troops seems to have occurred on Pharmouthi 20 (15 April), during the spectacle of Kraste (*thea tēs Krastēs*). On this day, there were 116 soldiers from outside Oxyrhynchus in the city;[106] presumably, these men were

---

97. Cf., e.g., Schmitt 1994 and Gascou 1976a.

98. Of course, I grant that *buccellarii,* like any armed force, might commit abuses (cf. the especially germane *W.Chr.* 471) and that their relationships with their *geouchoi* (though not governed by *paramonē* contracts; cf. p. 127 below) would have inevitably been those of patronage and dependency.

99. Alston (2002, 413) notes, "The texts from Egypt provide no clear answer."

100. See n. 52.

101. Lines 126ff. See p. 132 below.

102. Cf. Rémondon 1961, 65.

103. Gascou 1976a, 153–54 n. 4, has the fullest discussion.

104. Cf. *P.Oxy.* XVI 1921.16.

105. *P.Oxy.* XVI 1861.8–9.

106. From Kynopolis: table 4.6.7 (35); Herakleopolis: 4.6.31 + 4.6.33 (26); Koma: 4.6.27 (19); Kromidiotai: 4.6.16 + 4.6.17 (36).

TABLE 4.6. Military Distributions (*P.Oxy.* XXVII 2480)

Line(s)	Agent, Date	Recipient(s)	Amount	Notes
1. ll. 6–7	Phib *oinocheiristēs,* Thoth 9	27 *buccellarii* + *meizoteros* (probably Theodoros, cf. next entry)	7.25 *dipla* or .25 *d*/person w/.5 *d* for *meizoteros*	—λόγῳ ἀναλώματος —γεουχικῆς χρείας (= liturgy, so Gascou) ἕνεκα[a] —from Kynopolis to Oxyrhynchus
2. ll. 16–17	Phib *oinocheiristēs,* Thoth 10	27 recruits[b] + *meizoteros* Theodoros	7.25 *dipla* or .25 *d*/person w/.5 *d* for *meizoteros*	—λόγῳ ἀναλώματος —γεουχικῆς χρείας ἕνεκα —from Kynopolis to Oxyrhynchus
3. ll. 13–14	Phib *oinocheiristēs,* Thoth 11?[c]	27 *buccellarii* + *meizoteros* Theodoros	7.25 *dipla* or .25 *d*/person w/.5 *d* for *meizoteros*	—λόγῳ ἀναλώματος —γεουχικῆς χρείας ἕνεκα —from Kynopolis to Oxyrhynchus
4. ll. 3–4	Phib *oinocheiristēs,* Thoth 12	27 *buccellarii* + *meizoteros* Theodoros	7.25 *dipla* or .25 *d*/person w/.5 *d* for *meizoteros*	—λόγῳ ἀναλώματος —γεουχικῆς χρείας ἕνεκα —from Kynopolis to Oxyrhynchus
5. ll. 68–69	Phib *oinocheiristēs,* Pharmouthi 16	35 *buccellarii* with *meizoteros*	8.75 *dipla* or .25 *d*/person; nothing for *meizoteros*	—λόγῳ ἀναλώματος —from Kynopolis to Oxyrhynchus, ἕνεκα τῆς θέας τῆς Κράστης[d]
6. l. 75	Phib *oinocheiristēs,* Pharmouthi 17	35 *buccellarii*	8.75 *dipla* or .25 *d*/person	—λόγῳ ἀναλώματος —from Kynopolis to Oxyrhynchus, ἕνεκα τῆς θέας τῆς Κράστης
7. l. 80	Phib *oinocheiristēs,* Pharmouthi 18–20	35 *buccellarii*	36.25 *dipla* or .25 *d*/person/day	—λόγῳ ἀναλώματος —from Kynopolis to Oxyrhynchus, ἕνεκα τῆς θέας τῆς Κράστης
8. ll. 102–3	Phib *oinocheiristēs,* Pachon 30	20 recruits + *meizoteros*	5 *dipla* or .25 *d*/person; nothing for *meizoteros*	—λόγῳ ἀναλώματος —γεουχικῆς χρείας ἕνεκα —from Kynopolis to Oxyrhynchus

(continues)

TABLE 4.6—*(Continued)*

Line(s)	Agent, Date	Recipient(s)	Amount	Notes
9. l. 104	Phib *oinocheiristēs,* Pauni 11	20 recruits	5 *dipla* or .25 *d*/person	—λόγῳ ἀναλώματος —γεουχικῆς χρείας ἕνεκα —from Kynopolis to Oxyrhynchus
10. l. 50	Phib *oinocheiristēs,* Mecheir 29	30 Kromidiotai	30 *sextarii* = 5 *dipla* or 1 *sext.*/ person	—λόγῳ ἀναλώματος —to Oxyrhynchus; ἕνεκα τῆς ἐνορίας, "because of the *territorium*"
11. l. 51	Phib *oinocheiristēs,* Mecheir 29 (30?)	same 30 Kromidiotai, with *antigeouchos*	30 *sextarii* = 5 *dipla* or 1 *sext.*/ person; nothing for *antigeouchos*	—λόγῳ ἀναλώματος —to Oxyrhynchus
12. l. 52	Phib *oinocheiristēs,* Phamenoth 1?	30 Kromidiotai, with *antigeouchos*	30 *sextarii* = 5 *dipla* or 1 *sext.*/ person; nothing for *antigeouchos*	—λόγῳ ἀναλώματος —to Oxyrhynchus
13. ll. 59–60	Phib *oinocheiristēs,* Phamenoth 4	30 Kromidiotai, with *antigeouchos*	30 *sextarii* = 5 *dipla* or 1 *sext.*/ person; nothing for *antigeouchos*	—λόγῳ ἀναλώματος —to Oxyrhynchus; ἕνεκα τῆς ἐνορίας, "because of the *territorium*"
14. ll. 57–58	Phib *oinocheiristēs,* Phamenoth 6–7	30 Kromidiotai, with *antigeouchos*	60 *sextarii* = 10 *dipla* or 1 *sext.*/ person/day; nothing for *antigeouchos*	—λόγῳ ἀναλώματος —to Oxyrhynchus; ἕνεκα τῆς ἐνορίας, "because of the *territorium*"
15. ll. 53–54	Phib *oinocheiristēs,* Phamenoth 9–10	30 Kromidiotai, with *antigeouchos*	60 *sextarii* = 10 *dipla* or 1 *sext.*/ person/day; nothing for *antigeouchos*	—λόγῳ ἀναλώματος —to Oxyrhynchus; ἕνεκα τῆς ἐνορίας, "because of the *territorium*"
16. ll. 85–86	Phib *oinocheiristēs,* Pharmouthi 19–20	30 Kromidiotai	10 *dipla* or .167 *d*/person; if *dipla* of 6 *sext.*, then 1 *sext.*/person/day (as elsewhere)	—λόγῳ ἀναλώματος —to Oxyrhynchus; ἕνεκα τῆς ἐνορίας, "because of the *territorium*"
17. l. 91	Phib *oinocheiristēs,* Pharmouthi 20	another 6 Kromidiotai	1 *diploun* or .167 *d*/person; if *diploun* of 6 *sext.*, then 1 *sext.*/person (as elsewhere)	—λόγῳ ἀναλώματος —to Oxyrhynchus, ἕνεκα τῆς θέας τῆς Κράστης

TABLE 4.6—*(Continued)*

Line(s)	Agent, Date	Recipient(s)	Amount	Notes
18. l. 92	Phib *oinocheiristēs,* Pharmouthi 21	the same 36 Kromidiotai	6 *dipla* or .167 *d*/person; if *dipla* of 6 *sext.*, then 1 *sext.*/ person (as elsewhere)	—λόγῳ ἀναλώματος —to Oxyrhynchus, ἕνεκα τῆς θέας τῆς Κράστης
19. ll. 8–9	Phib *oinocheiristēs,* Thoth 16–17	8 *buccellarii*	4 *dipla,* 2 days or .25 *d*/person	—λόγῳ ἀναλώματος —ed.: "to regulate the tax-collecting district"[e] —to Kynopolis
20. l. 30	Phib *oinocheiristēs,* Phaophi 18	3 recruits	3 *dipla* for 2 days or .5 *d*/person	—λόγῳ ἀναλώματος —to Herakleopolis
21. l. 29	Phib *oinocheiristēs,* Phaophi 20	10 *buccellarii* + tribune Theodoros	3 *dipla* or .25 *d*/person w/.5 *d* for tribune	—λόγῳ ἀναλώματος —to *kōmē* Se(s)phtha (former 10th *pagus*)
22. l. 35	Phib *oinocheiristēs,* Hathyr	24 *buccellarii* with tribune	6 *dipla* or .25 *d*/person; nothing for tribune	—λόγῳ ἀναλώματος —for *exanysis* (exaction) of taxes —to *kōmē* Se(s)phtha
23. l. 49	Phib *oinocheiristēs,* Mecheir 28	9 *buccellarii*	2.25 *dipla* or .25 *d*/person	—λόγῳ ἀναλώματος —to Oxyrhynchus
24. ll. 55–56	Phib *oinocheiristēs,* Phamenoth 20	*buccellarii* (probably 2) with *antigeouchos*	.5 *diploun,* probably .25 *d*/person; nothing for *antigeouchos*	—λόγῳ ἀναλώματος —to Oxyrhynchus; ἕνεκα τῆς ἐνορίας, "because of the *territorium*"
25. ll. 66–67	Phib *oinocheiristēs,* Pharmouthi 16	19 *buccellarii* with tribune	4.75 *dipla* or .25 *d*/person; nothing for tribune	—λόγῳ ἀναλώματος —from Herakleopolite (Koma) to Oxyrhynchus, ἕνεκα τῆς θέας τῆς Κράστης
26. ll. 72–73	Phib *oinocheiristēs,* Pharmouthi 17	19 *buccellarii*	4.75 *dipla* or .25 *d*/person	—λόγῳ ἀναλώματος —from Herakleopolite (Koma) to Oxyrhynchus, ἕνεκα τῆς θέας τῆς Κράστης

(continues)

TABLE 4.6—*(Continued)*

Line(s)	Agent, Date	Recipient(s)	Amount	Notes
27. ll. 76–77	Phib *oinocheiristēs,* Pharmouthi 18–20	19 *buccellarii*	14.25 *dipla* or .25 *d*/person/day	—λόγῳ ἀναλώματος —from Herakleopolite (Koma) to Oxyrhynchus, ἕνεκα τῆς θέας τῆς Κράστης
28. ll. 64–65	Phib *oinocheiristēs,* Pharmouthi 15	18 *buccellarii*	4.5 *dipla* or .25 *d*/person	—λόγῳ ἀναλώματος —from Herakleopolis to Oxyrhynchus, ἕνεκα τῆς θέας τῆς Κράστης
29. ll. 70–71	Phib *oinocheiristēs,* Pharmouthi 16	18 recruits	4.5 *dipla* or .25 *d*/person	—λόγῳ ἀναλώματος —from Herakleopolis to Oxyrhynchus, ἕνεκα τῆς θέας τῆς Κράστης
30. l. 74	Phib *oinocheiristēs,* [Pharmouthi 17 (no date written)]	18 *buccellarii*	4.5 *dipla* or .25 *d*/person	—λόγῳ ἀναλώματος —from Herakleopolis to Oxyrhynchus, ἕνεκα τῆς θέας τῆς Κράστης
31. ll. 78–79	Phib *oinocheiristēs,* Pharmouthi 18–20	18 *buccellarii*	13.5 dipla or .25 *d*/person/day	—λόγῳ ἀναλώματος —from Herakleopolis to Oxyrhynchus, ἕνεκα τῆς θέας τῆς Κράστης
32. ll. 87–88	Phib *oinocheiristēs,* Pharmouthi 18	another 8 *buccellarii*	2 *dipla* or .25 *d*/person	—λόγῳ ἀναλώματος —from Herakleopolis to Oxyrhynchus, ἕνεκα τῆς θέας τῆς Κράστης
33. l. 89	Phib *oinocheiristēs,* Pharmouthi 19–20	the same 8 *buccellarii*	4 *dipla* or .25 *d*/person/day	—λόγῳ ἀναλώματος —from Herakleopolis to Oxyrhynchus, ἕνεκα τῆς θέας τῆς Κράστης
34. ll. 94–95	Phib *oinocheiristēs,* Pharmouthi 21	44 (*sic*—expect 45) *buccellarii*	5.5 *dipla* or .125 *d*/person; for a half of a day?	—λόγῳ ἀναλώματος —from Herakleopolis (25; 26 meant?) and Koma (Herakleopolite; 19) to Oxyrhynchus, ἕνεκα τῆς θέας τῆς Κράστης

TABLE 4.6—*(Continued)*

Line(s)	Agent, Date	Recipient(s)	Amount	Notes
35. l. 117	Phib *oinocheiristēs,* Epeiph 15	the same 16 *buccellarii,* cf. 4.6.37 and 38	6 *dipla* or .375 *d*/person[f]	—λόγῳ ἀναλώματος —from Koma and Herakleopolis *(sic?)* to Oxyrhynchus
36. ll. 109–11	Phib *oinocheiristēs,* Epeiph 16	52 recruits + 16 *buccellarii*	17 *dipla* or .25 *d*/person	—λόγῳ ἀναλώματος —to keep order at the *thea tou Pēchismou*[g] —from Kynopolis (32 recruits), Herakleopolis (16 *buccellarii*), and Koma (20 recruits), to Oxyrhynchus
37. ll. 112–14	Phib *oinocheiristēs,* Epeiph 18	32 recruits	9 *dipla* or .28 *d*/person (the 16 from Koma receive .31 *d*/person; those from Herakleopolis, .25 *d*/person).[f]	—λόγῳ ἀναλώματος —to keep order at the *thea tou Pēchismou* —from Koma (16 recruits) and Herakleopolis (16 recruits), to Oxyrhynchus
38. ll. 115–16	Phib *oinocheiristēs,* Epeiph 19	the same 32 recruits	9 *dipla* or .28 *d*/person (probably as previous: the 16 from Koma receive .31 *d*/person; those from Herakleopolis, .25 *d*/person).[f]	—λόγῳ ἀναλώματος —from Koma (16 recruits) and Herakleopolis (16 recruits), to Oxyrhynchus

[a]See p. 118.

[b]A photo of the papyrus reveals that the *delta* of the ed. pr.'s 24 (i.e., κδ) is overwritten by a *zeta.*

[c]Ed. pr.: Θῶθ ι̣[

[d]See p. 140.

[e]Greek: ἐπὶ τῷ καταστῆσαι τὸ πρόσωπον. In the text's commentary, *prosōpon* is defined as "the autopract district in which the landowner is directly liable for all taxes."

[f]For discussion, see p. 119.

[g]For this *thea,* see p. 140.

augmenting forces already there. None of the explicit missions of the troops seem to be "military" *sensu strictissimo.* They appear as police at festivals and in fiscal roles (e.g., assisting with collection; 4.6.22). Gascou has suggested that distributions tagged γεουχικῆς χρείας ἕνεκα indicate payments linked to "liturgical policing" incumbent upon the Apiones.[107] Those labeled "for the sake of the countryside" (ἕνεκα τῆς ἐνορίας/Latin *territorium*)—which all fall in February–March and, with one exception, involve Kromidiotai—are obscure. There is a mechanical quality to the language of 2480. It is doubtful that troops went back and forth between, say, Koma and Oxyrhynchus each day; when working the *thea tēs Krastēs,* for example, they certainly stayed overnight. All of the account's military entries are labeled λόγῳ ἀναλώματος; this should not be construed as "(reimbursement) for expenses" but, instead, has a sense closer to "allowance."[108]

The rations in 2480 are almost always one-quarter of a *diploun* per person per day. Since wine rations were typically either 1 or 2 *sextarii,*[109] and because the *diploun* hitherto has only been attested as a container of 4.5, 6, or 8 *sextarii,* it seems most likely that the *dipla* in 2480 contain 8 *sextarii* and, accordingly, that most of the rations in the account are 2 *sextarii.*[110] Unambiguous rations of 2 *sextarii* are common in contemporary military accounts from the Oxyrhynchite, and *buccellarii* are among their recipients.[111] Though it is true that several of the distributions to the Kromidiotai are explicitly in *dipla* of 6 *sextarii* (e.g., 4.6.15), it does not follow that the entire account is reckoned in this *diploun.*[112] The identification of the *dipla* in question as comprised of 6 *sextarii* occurs precisely because they were not the norm. *Dipla* of 6 *sextarii* are admittedly employed without specification (e.g., 4.6.16), but this occurs only with the Kromidiotai. These men, who received but a single *sextarius* each per day, may well have been on the same footing as the *symmachoi* (messengers) in *P.Oxy.* XVI 1920.6–7, that is, a lower grade of armed troop.[113] Table 4.6.20, which, on its face, suggests double rations for some *buccellarii,* may in fact reflect a situa-

107. Gascou 1985, 58. Sarris (2006, 172) argues that the label indicates private estate business.

108. Cf. Morelli 1996, 25–26.

109. Cf., e.g., Gascou 1989, 291. His calculation of the size of the wine ration for the *stratiōtai* in *P.Oxy.* XVI 1920 is, however, incorrect; see pp. 120–21, 123 below.

110. It may be that a ration of this size was not consumed on a continual basis by the soldier alone; cf. the figures in Clarysse and Vandorpe 1997, 71 and n. 10, along with the remarks in Foxhall and Forbes 1982, 73. Also note pp. 132–33 below.

111. See table 4.8, e.g.

112. Pace Morelli (1996, 14 n. 7), who is followed by Mitthof (2001, 243).

113. For the *symmachoi,* see Jördens 1986a and 1992a. If the date in table 4.6.11 is correct, the Kromidiotai did receive a full (i.e., standard) ration on Mecheir 29 (4.6.10 + 4.6.11).

tion similar to the one that seems to be operative in 4.6.35–38. In 4.6.37 of Epeiph 18 (and no doubt also in 4.6.38 of Epeiph 19), sixteen recruits from Koma are said to have received higher rations than their counterparts from Herakleopolis. The account probably indicates—the reference to Herakleopolis is likely a scribal slip—that the same troops from Koma also drew larger rations on the fifteenth (4.6.35), while a larger contingent (twenty men) from this locale received the standard ration on the sixteenth (4.6.36). There is no distribution recorded for the seventeenth, and this omission—if the sixteen recruits were among those receiving the standard ration on the sixteenth—suggests a solution: the wine for the seventeenth is in the entries for the other three days (16 *dipla* [from three days: 4.4.35, 37–38] ÷ 64 man-days = .25 *d*/man-day, the standard ration). If this interpretation is correct, either the accounts reflect reality (i.e., no distribution on the seventeenth), or the *oinocheiristēs* has doctored them to cover missing documentation or some other lapse.[114]

Morelli has noted that the connector (καί, σύν, or μετά) employed by the *oinocheiristēs* in those entries in which an individual accompanies troops indicates whether or not this person also received wine (always a double ration).[115] With σύν and μετά, no distributions are made, and Morelli states that in these cases, the accompanying individuals were mentioned solely "to identify the group of beneficiaries." There may be truth to this, but the suggestion is not compelling: the hypothesis is applicable only to Pharmouthi 16 (4.6.5, 25, 29), and on this date, a unit's geographic origin alone would have sufficed to identify it. In any case, it is striking that the recipient of the double ration is always named Theodoros.[116] In 4.6.1–4, the Theodoros in question is a *meizoteros;* in 4.6.21, he is a tribune. It seems likely that these entries all concern the same individual.[117] If this is correct, an explanation is furnished for Theodoros' special treatment: he was probably the equivalent of the literary sources' *maior domus,* a high-level commander of *buccellarii*—in this case, the troops associated with the estate.[118]

---

114. The account, a clean copy, is certainly retrospective.

115. Morelli 1996, 14 n. 7, correcting Fikhman 1968, 131.

116. In table 4.6.1 a name is not given, but it is almost certain that Theodoros was meant.

117. Cf. already Fikhman 1968, 132 (stated far too tentatively).

118. For the *maior domus,* see Schmitt 1994, 164–65 and refs. Gascou (1976a, 147 n. 2, 152) has suggested the equivalence with *meizoteros* (rejected by Schmitt). Sarris 2006, 76, does not mention a military role for the *meizoteros:* "As with all the high-ranking employees, the *meizoteros* operated within a wide sphere of administrative responsibility, but was primarily in charge of the maintenance of the central household's stores and supplies, and the provisioning in kind of the workforce and his fellow administrators." *SB* XXII 15368, in which the *lamprotatos* Alexandros receives half a *knidion* per day "for my allowance" (ὑπὲρ ἀναλώματός μου), no doubt concerns another military *meizoteros.* Accordingly, the *knidia* in this text held 8 *sextarii.*

P.Oxy. XVI 1920 is not an *oinocheiristēs'* account and does not solely concern wine distributions. Connected with the arrival of "the men of the most extraordinary *patricius* Athanasios" (l .1)[119]—both soldiers and *cohortales*—from the Thebaid, it covers all of the standard military rations (meat, wine, bread, and oil), as well the *capitum* (fodder ration) and, in most cases, wood. (*Cohortales,* members of the civil service, the *militia officialis,* also drew the *annona.*) Table 4.7 lists the wine distributions in the account. Though lines 16–17 of the text assign monetary values to the rations, it seems likely that supplies were actually distributed, that this was not a case of commutation. The level of detail in the records seems excessive for commutation; moreover, the recipients were away from their base in the Thebaid.[120] It seems likely that the sums of *solidi* listed simply served as a record of the credits that the Apiones could claim against their tax.[121]

The soldiers in table 4.7.1 are elite "regulars," the Scythae Iustiniani.[122] Calculating their ration is problematic. The text (ll. 4–5) reads, "83 2/3 *sextarii* in their *sextarius,* the equivalent of 167 1/3 *sextarii,* with 8 *sextarii* to the *diploun,* total 20 ½ ¼ 1/8 *dipla.*"[123] The Scythians' *sextarius* is thus twice as large as the bookkeeper's measure, but there are two ways of interpreting this: either the bookkeeper's *sextarius* is "standard," or the soldiers' is. The former interpretation might seem less violent. Under it, only the soldiers' *sextarius* is abnormal (= 2 "regular" *sext.*), and this could simply be the bookkeeper's way of indicating a double ration. Yet the resulting ration of 4 *sextarii* is massive, four times that received by the Scythians at home[124] and double that of the *buccellarii* in 2480. Unless there were extenuating circumstances, consistent largesse of this sort seems unlikely. The other interpretation requires the bookkeeper's employment of a half-*sextarius* measure. Under it, the soldiers' ration now corresponds with other testimonia,[125] but one must grant a *diploun* with a capacity of 4 *sextarii,* and this size vessel is otherwise unattested. The half-*sextarius* measure is also troubling because the bookkeeper employs the stan-

119. For Athanasios, see Morelli 2008; Morelli's pp. 228–30 concern 1920.

120. Cf. Haldon 1984, 112.

121. Cf. *NJ* CXXX.

122. For this unit, see Gascou 1989, 309–10, and Mitthof 2008 (in the context of Justinian's military reforms in the Thebaid).

123. οἴνου ξέσται πγω ἐκ τοῦ ξέστου αὐτῶν εἰς ξέστας ρξζγ', τῶν ξεστῶν η διπλοῦ α, διπλᾶ κLdη'.

124. Cf. *SB* XX 14494.27.

125. Cf. *P.Oxy.* XVI 2046 (table 4.8), which also indicates (at l. 43, e.g.) that an *annona* contained 2 *sextarii* of wine and that the regular soldier received 1 *annona* (likewise implied by the schedule in *CJ* I.27.2.19–34).

TABLE 4.7. Military Distributions (*P.Oxy.* XVI 1920)

Lines	Agent, Date	Recipients	Amount	Notes
1. ll. 3–5	?; Mecheir 3–13	41.833 *stratiōtai tōn Skythōn*	total daily 83.667 *sextarii* (standard) = 167.333 *sext.* (halves); in *dipla* of 8 *sext.* (halves) = 20.875 *d* (*sic;* = 20.917 *d*); 12 days = 250.5 *d* total (actually 251 *d*); = .5 *d* (half-*sext.*)/man/day	See pp. 120–21, 123.
2. ll. 6–7	?; Mecheir 3–12	58 *symmachoi*	total daily 58 *sext.* = 9.667 *d,* therefore *d* = 6 *sext.;* total 106.333 *d* = .167 *d*/man/day or 1 *sext.*	
3. ll. 8–10	?; Mecheir 3–13	*cancellarius,*[a] *cursores,*[b] *praecones,*[c] and others; total 24	total daily 48 *sext.* = 8 *d,* therefore *d* = 6 *sext.;* total 88 *d* = .333 *d*/man/day or 2 *sext.*	
4. ll. 11–12	?; Mecheir 4–13	30 *symmachoi* of the *riparii*	total daily 15 *sextarii* = 2.5 *d,* therefore *d* = 6 *sext.;* total 25 *d* = .083 *d*/man/day or .5 *sext.*	

Note: In line 13 of this document, the total number of *dipla* appears, and no accounting is made for the different-sized vessels that were used. Likewise, in the distributions for the following two days (Mecheir 14–15, ll. 14–15), which are recorded as lump sums, no mention is made of the varying capacity of the vessels. It was the number of jars, then, that was being tallied and not the volume of wine *sensu stricto;* the counting of gold pieces by tale (*arithmia nomismatia*) seems a good parallel.

[a]See Rouillard 1928, 46, Jones 1964, 602–3, and *P.Eirene* II 8, 5n. (See now the thorough discussion in *APF* 55 [2009] 56–62.)

[b]See *P.Oxy.* LVIII 3934, 9n.; *P.Harrauer* 53, 5n.; *P.Horak* 33, 3n.

[c]See Jones 1964, 593.

dard *sextarius* elsewhere in the account. Nevertheless, it seems that the half-*sextarius* interpretation is the correct one. The *oinokreon* commutation values (pounds of meat + *sext.* of wine/*sol.*) are identical in 4.7.2 and 4.7.3 (138 units/*sol.*).[126] A 4-*sextarius diploun* in 4.7.1 results in a commutation value of 138 for that entry—a perfect match.[127] Furthermore, while a *diploun* of 4 *sextarii* is unknown, there is a single attestation of a 4.5-*sextarius* "double," and

126. Using the commutation tariffs in lines 16–17 (and assuming a *diploun* of 6 *sext.*—the standard in ll. 7, 10, 12, and 18—in l. 17). Table 4.7.4 is discrepant: 133 units per *solidus.* For *oinokreon* and its value, see Mitthof 2001, 209–11, 262–63.

127. If the bookkeeper's *sextarius* is the standard one, 4.7.1's *oinokreon* commutation value is 143.

TABLE 4.8. Military Distributions (*P.Oxy.* XVI 2046)

Line(s)	Agent, Date	Recipient(s)	Amount	Notes
1. l. 3	?; Epeiph 4	26 *buccellarii*	6.5 *dipla* or .25 *d*/man	
2. l. 4	?; Epeiph 4	14 "Scythians"	28 *sextarii* or 2 *sext.*/man	
3. l. 5	?; Epeiph 4	tribune Oulior	2 *dipla*	
4. l. 6	?; Epeiph 4	4 *paidaria* of tribune[a]	8 *sextarii* or 2 *sext.*/man	
5. l. 9	?; Epeiph 5	26 *buccellarii*	6.5 *dipla* or .25 *d*/man	
6. l. 10	?; Epeiph 5	14 "Scythians"	28 *sextarii* or 2 *sext.*/man	
7. l. 11	?; Epeiph 5	the other 20 "Scythians"	40 *sextarii* or 2 *sext.*/man	
8. l. 12	?; Epeiph 5	these Scythians' *paidaria*	4 *sextarii* or 1 *sext.*/man	
9. l. 13	?; Epeiph 5	tribune Oulior	2 *dipla*	
10. l. 14	?; Epeiph 5	4 *paidaria* of tribune	8 *sextarii* or 2 *sext.*/man	
11. ll. 17–25	?; Epeiph [6–]10	4 Scythians + 9 *buccellarii*	26 *sextarii*/13 men/day or 2 *sext.*/man/day	ll. 17–18 (+ *BL* XI 158): τοῖς οὖσ(ιν) ἐνταῦθ(α) ἀηδι̣[ῶν ἕνεκα ? τῶν στρα(τιωτῶν)] τῶν Σκυθῶν οὕτ(ως)·; eds.: "presence . . . due to some disturbance."
12. l. 41	?; Pauni 23	7 men: 3 *buccellarii,* 1 *cancellarius,* 1 *cursor,* 2 *ostiarii*[b]	[14 *sextarii,* or 2 *sext.*/man/day]	
13. ll. 42–43	?, Pauni 24	36 Scythians = 43 *annonae;* their 14 *paidaria* = 7 *annonae;* for 7 *contubernales,*[c] 4 *annonae*	total 54 *annonae* = 108 *sextarii;* each *annona* = 2 *sextarii*	
14. ll. 44–45	?; Pauni 24	29 *buccellarii* = 30 *annonae;* 8 *paidaria* = 4 *annonae*[d]	total 34 *annonae* = 64 *sextarii;* each *annona* = 2 *sextarii*	
15. l. 55	?; Pauni 24	4 men: 1 *cancellarius,* 1 *cursor,* 2 *ostiarii*[b]	5 *annonae* = 1.25 *dipla* or .25 *d*/*annona;* since *annona* = 2 *sextarii,* *d* = 8 *sextarii*	

TABLE 4.8—*Notes*

[a]For these *paidaria,* see Gascou and Worp 1990, 223, and Rémondon 1961, 61. The latter excludes slaves from the picture altogether, in contrast to Zuckerman 1988, 284, who admittedly is concerned with the fourth century but does cite the *Strategikon.*

[b]*Ostiarii:* See *P.Horak* 56, 3n. and refs. For the *cancellarius* and *cursor,* see table 4.7, nn. a and b, respectively.

[c]In general, see Rémondon 1961, 72, and Haldon 1984, 109 and 382, where the relevant passages in the *Strategikon* are cited. Also cf. next note.

[d]Breakdown: (1) *Contubern(ales)*—so ed. pr., but *contubern(ium)* is more appropriate if, as is probable here, the named figure is also a recipient of rations—of Zemarchos, 7 *buccellarii* = 8 *annonae* + 3 *paidaria* = 1.5 *annonae;* (2) *Con.* Boraïdou, 4 *buccellarii* = 4 *annonae* + 1 *paidarion* = .5 *annona;* (3) *Con.* Georgiou, 2 *buccellarii* = 2 *annonae* + 1 *paidarion* = .5 *annona;* (4) *Con.* Paulou, 6 *buccellarii* = 6 *annonae* + 1 *paidarion* = .5 *annona;* (5) *Con.* Iordannou (cf. *BL* VI 105), 2 *buccellarii* = 2 *annonae;* (6) *Con.* Alexandrou, 4 *buccellarii* = 4 *annonae* + 1 *paidarion* = .5 *annona;* (7) *Con.* Martinou, 1 *buccellarius* = 1 *annona,* 1 *paidarion* = .5 *annona;* (8) *Con.* Kyrikou, 3 *buccellarii* = 3 *annonae.*

it appears in connection with the Scythians.[128] We likely encounter these lower-capacity vessels (2.2–2.5 liters) because it was difficult or impractical for a soldier to carry around more liquid weight.[129] Their specificity or, rather, their infrequent appearance may have led the bookkeeper to describe the 4-*sextarius* version in terms of a vessel commonly used on his estate for military rations (and many other purposes), the *diploun* of 8 *sextarii.* The soldiers' wine was probably decanted, in fact, from such jars.

P.Oxy. XVI 2046 is quite similar to 1920 and dates to just five months later (on the assumption that the fiscal indiction was being employed in these accounts).[130] Table 4.8 lists its wine distributions. On the recto (ll. 1–39, table 4.8.1–11), wine rations for the regulars—with the exception of their tribune—and for those accompanying them are always in *sextarii.* Soldiers each receive 2 *sextarii,* while their "boys" (*paidaria;* except those of the tribune) receive half of this.[131] The verso of the text (ll. 40–58, table 4.8.12–15) introduces the *annona* as a unit for rations. The wine component of each *annona* unit is 2 *sextarii.* The standard military distribution was no doubt one *annona* per recipient; the extras in 4.8.13–15 presumably went to officers and other senior figures.[132] A *paidarion* received half an *annona;* the *contubernales* (literally, "tent companions") in 4.8.13, somewhat less than this.

128. *P.Oxy.* XVI 2046.7.

129. I suspect that the soldiers' vessels were, in fact, wineskins.

130. Otherwise, to the following year, as Morelli (*BL* X 146) dates it. Columns i and ii of the recto and column iii and the verso are separate writing acts in distinct hands. (Actually, pace the ed. pr., I believe that col. iii of the recto is in a third hand.) Sense suggests that the first two columns of the recto were written last.

131. For the precise meaning of *paidarion* here, cf. table 4.8, note a.

132. Cf. the schedule in *CJ* I.27.2.19–34. In table 4.8.14, the extra *annona* no doubt went to the troop leader; in 4.8.13, the situation is less clear, but cf. the comments in Zuckerman 1988, 282–85.

All but one of the ration entries for *buccellarii*, as well as those for the tribune, are given in *dipla*. Lines 7 and 15 (both read "8 ½ *dipla* of wine and 36 *sextarii*, the equivalent of 8 *dipla*," οἴνου διπλᾶ ηL καὶ ξέσται λς εἰς διπλᾶ η) and line 57 ("1 ¼ *dipla* of wine and 176 *sextarii*, the equivalent of 36 *dipla* plus ½ *sextarius*, total 40 1/3 *dipla*," οἴνου διπλοῦν αδ′ καὶ ξέσται ϱος εἰς διπλᾶ λθ καὶ ξέστου L, διπλᾶ μγ′) suggest a *diploun* of 4.5 *sextarii*, but pace Morelli,[133] it is likely that not all of the *dipla* in this text were the same size. Those of the tribune Ulior were probably like those of the Scythians, that is, the smaller vessels of 4.5 *sextarii*; 2 *dipla* would thus give him the double ration attested for the *meizoteros/tribune* in, for example, table 4.6.1–4 and table 4.6.21. The *dipla* of the *buccellarii*, however, probably held 8 *sextarii*; this capacity would give them a two-*sextarius* ration, which is both the equivalent of that given to the Scythians (cf. table 4.8.11) and a quantity attested for them in other texts. An 8-*sextarius* vessel furthermore results in a standard *oinokreon* (*sextarii* of wine:pounds of meat) ratio (2:1).[134] The large *diploun* might seem out of place for soldiers, but if the *buccellarii* were local troops, it probably did not pose significant difficulties.[135]

Morelli incorrectly asserts that there were "two different types of ἀννῶναι: one for the military, the other for officials."[136] In table 4.8.15, the officials are simply receiving the larger (8-*sext.*) vessels; after table 4.7.3, one expects that members of the *militia officialis* would have rations corresponding to those of the soldiers.[137]

Table 4.9 summarizes the military distributions in *PSI* VIII 953.[138] The data are familiar. Most often encountered are *buccellarii* receiving one-quarter *diploun* per day. This ration is also attested in *P.Oxy.* XVI 2046 and *P.Oxy.* XXVII 2480, and the arguments for an 8-*sextarius* vessel in the latter text also apply to 953. Notable in 953, besides the *buccellarius* associated with Ravenna (4.9.5), are the apparent distributions to women,[139] who may have received

133. Morelli 1996, 69.

134. Cf. Gascou 1989, 292.

135. Local troops: i.e., picked up along the way (from, say, Kynopolis) by Ulior. The account heading, incidentally, seems to refer to all of the troops, including the Scythians, as *buccellarii*. For the rise of "regular" *buccellarii*, see Haldon 1984, 101–2.

136. Morelli 1996, 72.

137. As was also the case in *P.Oxy.* 1920 (see table 4.7 notes), the totals of *dipla* given in 2046 (e.g., at l. 7) are records of quantities of *jars*, not capacities. This observation also applies to the stewards' accounts, but because vessels on the estate standard (for such, cf. *P.Oxy.* LVIII 3942.18–19) were generally employed, estimating the volume of wine distributed would not have been difficult.

138. I have excluded the entries in the fragments of *PSI* VIII 956 (and 955), all of which require at least some speculation.

139. Cf. Rémondon 1961, 61; Alston 2002, 415 (and refs.).

Table 4.9: Military Distributions (*PSI* VIII 953)

Line(s)	Steward/Date	Recipient(s)	Amount	Notes
1. l. 18	Serenos/Mesore	*buccellarii*	11 *solidi* less 44 *keratia* = 11 *sol.* private standard	—*paramenontes* τῷ ἐνδόξῳ οἴκῳ —sum perhaps brought by some *symmachoi,* for the same 1st indiction
2. ll. 21–23	Serenos/Mecheir or Phamenoth?	soldiers, presumably, of *archōn* (governor) Florentius	12 *sol.* less 48 *ker.* = 12 *sol.* private standard	money sent to village of Spania ὑπὲρ̣ ἀναλ[(ώματος) τοῖς ἐλ]θο̣ῦ̣σ̣(ι) εἰς τὴν ἐνορ̣ίαν (*territorium*)
3. l. 29	Serenos/ Phamenoth	Markianos s. Ioannes	7.5 *dipla* or .25 *d* for 30 days	—*paramenōn* τῷ ἐνδόξῳ οἴκῳ —λόγῳ ἀναλώματος
4. ll. 33–34	Serenos/ Pharmouthi 15–30	Epiphanios and Ioannes, *buccellarii*	8 *dipla* over 16 days or .25 *d*/man/day	—*paramenontes* τῷ ἐνδόξῳ οἴκῳ —λόγῳ ἀναλώματος
5. ll. 35–36	Serenos/ Pharmouthi 3–30	the other three *buccellarii:* Martinos *tōn* Rabenneson; Ioannes Belkidiou; Martinos Sabinianou	21 *dipla* over 28 days or .25 *d*/man/day	—*paramenontes* τῷ ἐνδόξῳ οἴκῳ
6. ll. 38–39	Serenos/ Pharmouthi 30 —Pachon [3?][a]	13 *buccellarii,* 4 women ?, and 9 *paidaria*	21.5 *dipla:* *buccellarii* no doubt receive .25 *d*/man/day; if over 4 days, and if *paidaria* receive .125 *d*/man/day (cf. table 4.8.8),[b] then women receive ration equivalent to *buccellarii.*	—coming *heneka?* (ed. pr.: ἐν [....]) *tou peribleptou Papiriou* —λόγῳ ἀναλώματος
7. l. 41	Serenos/?	? *buccellarii* from Kynopolis	5.25 *dipla;* if for one day, rations for 21 troops	—coming to Oxyrhynchite with *geouch(ikos) apokris(iarios)?*[c]

(continues)

Table 4.9—*(Continued)*

Line(s)	Steward/Date	Recipient(s)	Amount	Notes
8. ll. 43–44	Serenos/ Phamenoth 16–30	Pa–,[d] s. Ioustos, and Theodoros, s. Andreas, *buccellarii*	7.5 *dipla*[e] or .25 *d*/man/day	—*paramenontes* τῷ ἐνδόξῳ οἴκῳ —λόγῳ ἀναλώματος
9. l. 45	Serenos/Pachon 23–30	K–, *buccellarius*	2 *dipla* or .25 *d*/day	—*paramenōn* τῷ ἐνδόξῳ οἴκῳ —λόγῳ ἀναλώματος
10. ll. 48–49	Serenos/Pauni	*buccellarii*	302.5 *dipla* = 40.333 daily rations of .25 *d*	—*paramenontes* τῷ ἐνδόξῳ οἴκῳ —λόγῳ ἀναλώματος —in accordance with "parchment of expenses of the 1st indiction"
11. l. 51	Serenos/Pauni (presumably)	*buccellarii* (v. probably) with someone described as *endoxotatos illoustrios* (prob. pagarch)	4 *dipla* or 16 .25 *d* rations	—coming to Oxy-rhynchus
12. ll. 68–69	Serenos/Pachon 24 (for 30 days)	10 *buccellarii*	[75 *dipla*]—if standard rations	—escorting 2nd installment of tax revenue to Alexandria —λόγῳ ἀναλώματος
13. ll. 78–79	Serenos/30 days	at standard rate, 12 (= 90 *dipla*) or 13 (= 97.5 *dipla*) *buccellarii*	90 ≤ *dipla* ≤ 99	—escorting [1st] installment of tax revenue to Alexandria[f] —λόγῳ ἀναλώματος

[a]There is a lacuna (±10 letters) following "and (Pachon) 2." After said καὶ β̄, perhaps read καὶ γ̄, followed by ἡμερ(ῶν) δ.

[b]Of course, it is possible that these *paidaria* are children; for distributions to children, cf. p. 132.

[c]Assuming that ἐλθοῦ[σι—the sigma in the ed. pr. is no longer extant—was followed by ἐνταῦθα μετὰ τοῦ *vel sim.*

[d]Pa[might also be the heading for the month, i.e., Πα[χών.

[e]The 3.5 (= γL) of the ed. pr., read uncertainly, is incorrect.

[f]The surrounding dates (in Tybi and Mecheir) suggest that we are dealing with the first installment (*katabolē*), which was due on the Kalends of January, cf. *CJ* X.16.13. (For additional discussion of *katabolai,* see Zuckerman 2004, 182–84.)

rations equivalent to those of the soldiers (4.9.6).[140] There are also the payments in coin (4.9.1, 2), no doubt instances of *adaeratio* (commutation of rations in kind into money).[141] Finally, the frequent usage of the present participle of *paramenō* ("remain") in descriptions of the *buccellarii* in this text (4.9.1, 3–5, 8–10) has led Gascou to suggest that these men were governed by *paramonē* agreements, an argument recently adopted and expanded by Sarris.[142] Schmitt has rejected Gascou's position,[143] while Mazza has criticized Sarris—justifiably so.[144] These participles clearly do not have a technical sense; they merely indicate *buccellarii* from other sites (e.g., Herakleopolis or Koma) who are staying at Oxyrhynchus.[145]

Another estate account with military rations, *P.Oxy.* XVIII 2196r (after 587), is puzzling on its face.[146] It has the opening of a *pronoētēs*' account (and a known *pronoētēs*), but only the total of the *lēmmata* is recorded.[147] This is followed by a summary of the expenditures for the churches in the *prostasia* and then, curiously, by a list of payments—clearly the *annona militaris*—for the period Epeiph 22–29.[148] These payments include, in addition to bread (150 *litrai*/day), meat (50 *litrai*/day, commuted to *folles*, the base coinage), and oil (6.25 *sext.*/day), 17 *dipla* of wine per day, or, for the entire eight-day period, 136 *knidia* (*sic*—note the equivalence). If these are *dipla* of 8 *sextarii* and if the rations are the typical 2 *sextarii*, then the number of men, presumably soldiers, was sixty-eight. Such a unit strength, however, does not correlate neatly with the meat ration,[149] and since the daily bread ration for soldiers in estate accounts is typically 3 *litrai*,[150] a troop of fifty men and *dipla*/*knidia* of 6 *sextarii*, with an officer probably getting a double ration of wine, seems likely.[151]

---

140. See table 4.9.6 for details, and also note the discussion of household distributions below.

141. In *PSI* 953.23, the *solidi* expended are described as ἐνεχθέντα εἰς τὸ λῆμμα τούτου τοῦ λόγου ("applied to the income of this account"), a transfer to offset the abnormal payment in coin in the *analōmata;* cf. the discussion of *PSI* VIII 954.56–57 above (p. 78 n. 94).

142. Gascou 1976a, 145; Sarris 2006, 167–72. For *paramonē* agreements, see p. 129 below.

143. Schmitt 1994, 160 n. 100. Gascou, however, maintains his stance in the 2008 version of his text.

144. Mazza 2008, 154.

145. *PSI* VIII 953 does have maintenance payments for "permanent" resident *buccellarii;* cf. p. 131 below.

146. For the date, see *BL* VIII 255 and XII 144, as well as p. 16 n. 103 above.

147. Incidentally, this sum suggests a smaller *prostasia* than those known from detailed accounts.

148. The *annona* was not recognized by the editor, but it did not escape the notice of, e.g., Jones (1964, 1262 n. 44).

149. In estate documents, one expects a meat ration of 1 or 2 pounds; cf. Mitthof 2001, 243–44.

150. Cf. Mitthof 2001, 243–44.

151. Were these Kromidiotai? Cf. p. 118 (with n. 113) above.

The strange location of these entries requires explanation.[152] All other evidence indicates that the *oinocheiristēs* was the sole distributor of annonary wine, and this was no doubt also the case for the *choiromageiros* (pork butcher) and meat, and so forth.[153] Distributions by a *pronoētēs* should give pause. The editors of 2196 note that it is "probably only a draft" (p. 167), and this comment suggests a plausible explanation: the papyrus was a piece of scrap that was used for scribal practice or the like.[154]

In addition to the estate accounts, two other texts provide testimonia that merit discussion. The first, *P.Oxy.* I 150 (590), is a receipt for a distribution to fourteen *buccellarii* from Kynopolis and Herakleopolis who had come to Oxyrhynchus on account of the *machē* ("fight"?) of Kosmas *apo meizoterōn* (retired *meizoteros*).[155] The text concerns a single day's rations, which total 3.5 *knidia*, or .25 *knidion* per man. These *knidia* are no doubt equivalent to the *dipla* in *P.Oxy.* XXVII 2480 and *PSI* VIII 953; that is, they each contain 8 *sextarii*, the capacity that yields the ubiquitous individual ration of 2 *sextarii*. The other text, *SB* XVI 12608, dates to 511 and has no explicit link to the Apiones;[156] indeed, it might not even concern armed men. In this papyrus, Ouiktor the *oinocheiristēs* distributes wine to a certain number of individuals who had come to Oxyrhynchus to assist, it seems, with the collection of a substantial sum of money owed to an *oinopratēs* (wine merchant).[157] The editor initially read τοῖς [. . βουκκελλα]ϱ(ίοις) as the recipients; this was later amended to τοῖς π̣[ϱ](εσβυτέϱοις) (*BL* IX 287), but the correct reading is in fact τοῖς π̣[αιδα]ϱίοις.[158] *Paidarion* is a vexed term (see the next section), but it can "admit . . . military, paramilitary, and police senses."[159] One *diploun* per day

152. Morelli 1996, 78, is unconvincing: "The fact that the expenses of grain, wine and oil have no direct connection with the income does not create difficulties, since this account records only income and expenditure but lacks a final balance."

153. *Choiromageiros:* See *P.Oxy.* LXXII 4923–25.

154. The editio princeps does not indicate any changes of hand, but comparison of lines 1–6 and 11–13 on a digital image suggests that they were separate acts of writing (at the least). For practice by Apion scribes, see Harrauer and Sijpesteijn 1986 (on *P.Princ.* II 96). *P.Oxy.* XVI 1910 is another text (a *pronoētēs'* account) that served as a model, but in both of these cases, practice occurred on the exemplar papyrus itself. A careful study of the handwriting of the Apion dossier is a desideratum. For the verso of 2196, see Hickey 2008.

155. Cf. *BL* VI 95; the text simply has ἀπὸ μειζ( ). Was this *machē* a disturbance requiring police intervention?

156. The same wine steward appears, however, in the unpublished P.Wash.Univ. inv. 251, the content of which is highly suggestive of the Apiones. It should be noted that 12608 dates to the time of Apion I's exile; for discussion of this period, see p. 11 above.

157. Cf. *BL* IX 287.

158. After a photo. The reading was first proposed to me by Nick Gonis.

159. Gascou and Worp 1990, 223–24.

(for a total of nine days) was distributed. If the *dipla* were of 8 *sextarii,* and if the individual ration consisted of a single *sextarius* (cf. table 4.8.8), the group would have consisted of eight men.

Labor

Two entries in *P.Oxy.* LVIII 3960 concern payments made to those in the service of the *oikos.* The first (l. 27) reads, "To the free *paramonarioi* for their 'salaries' for the ninth indiction, 8,906 *knidia* of wine";[160] this quantity of wine amounts to almost 27 percent of the steward's *hypodochē. Paramonarios* is, of course, suggestive of the well-known *paramonē* contracts of the Ptolemaic and Roman periods.[161] Under the strictest interpretation, these were effectively antichretic loans, with the debtor's labor over a period serving to pay the interest (and, exceptionally, the principal); in return, the debtor received a maintenance payment (consisting of food, clothing, etc.). Though Hengstl, for example, states, "It looks as if *paramonē* and the service contract flow into one another in the Byzantine period,"[162] agreements with the essential features of *paramonē* clearly still existed in the post-Diocletianic era.[163] Nevertheless, *paramonarios*—a rare word possibly attested in more Coptic texts than Greek ones[164]—seems to refer never to individuals under such antichretic contracts but, rather, to those under "simple" service agreements.[165] This no doubt holds true for the *paramonarioi* in 3960.

The *metrēmatiaioi,* a group of permanent workers contractually bound to the third-century Appianus estate, would seem to furnish a suitable comparandum for the Apiones' *paramonarioi.*[166] It seems unlikely that lower-level managers like the *pronoētai* are to be classified among their number; though it would be a mistake to discount the importance of patronage in the lives of

---

160. τοῖς παραμοναρίοις ἐλευθέροις ὑπὲρ ὀψωνίων θ ἰνδικτίονος οἴνου κνίδια ͵ηϡϛ.

161. See, e.g., Adams 1964; Samuel 1965; and Hengstl 1972, 9–34.

162. Hengstl 1972, 12.

163. Cf. *P.Heid.* V, pp. 287–88, where Jördens summarizes, "Doubtless the legal arrangement usually called *paramonē* has also survived still in the Byzantine period."

164. Coptic: *CPR* IV 160, 161, 168a (a complete edition of the last of these appears in Förster 1998, 290–92, which also has a correction to 160). These are instances of the loanword; ⲣⲙⲛϩⲱⲧⲣ, "hireling," has been suggested by Till as the Egyptian equivalent (Förster, p. 292). Greek: *P.Oxy.* LVIII 3960, *SB* I 4490, and *SB* XVI 13016, where the editors have proposed it for a lacuna (l. 16; the endorsement has *homologia paramonēs*). The reference from the Apion dossier is probably the earliest.

165. Pace Sarris 2006, 65: "workers of a non-servile condition bound by paramonar contracts."

166. For the *metrēmatiaioi,* see Rathbone 1991, 116–47 (whose interpretation of *paramonē* differs from the narrow one given above).

Apion managers,[167] the relationship of the *paramonarioi* with the *oikos* was likely one of greater dependency. Still, there are similarities between the two groups. It has already been suggested that the *paramonarioi* were governed by contracts,[168] and this was certainly the case for managers like the *pronoētai.*[169] The lengths of these contracts could even have been the same.[170] Like *paramonarioi,* managers were also recipients of *opsōnia* (salaries),[171] though it should be noted that remuneration terminology is tricky during the period under consideration. In general, *opsōnia* are considered to indicate a permanent employment relationship, while *misthoi* (wages)—as a rule of thumb, more narrowly referring to cash payments only—suggest casual labor.[172] There are, however, exceptions,[173] and one of the *pronoētēs* contracts, in fact, describes the steward's remuneration as μισθὸν ἤτοι [;[174] in the lacuna, Rea has reasonably supplied ὀψώνιον.[175]

Line 28 of *P.Oxy.* LVIII 3960 reads, "To the Egyptian *paidaria* for their 'salaries' for the ninth indiction, _348 *knidia* of wine";[176] the thousands place could have been occupied by a numeral as high as 6 (͵ϛ), but it should also be noted that additional distribution entries might have followed after the text breaks here.[177] The status of a *pais, paidarion,* or the like is often a difficult question. In this case, the fact that the *paramonarioi* in the preceding entry are called *eleutheroi* might be taken to mean that the *paidaria,* in contrast, were servile, but pace the text's editor Rea, it seems certain that these "chil-

---

167. In the case of the *pronoētai,* their remuneration from the *oikos* itself was considerably less than the "entry fee" that they paid for the position; cf. *P.Oxy.* LV 3804, 154n.

168. Also note that, with the exception of *P.Oxy.* LVIII 3960, the texts listed in n. 164 are contracts.

169. For the contracts of *pronoētai,* see Benaissa 2007.

170. The terms of *P.Oxy.* I 136 and LVIII 3952, both of which concern estate *pronoētai,* are one year. Of the extant agreements concerning *paramonarioi* (none of which derive from the estate), only *CPR* IV 160 seems open-ended; the others have durations of one year. It has been suggested, however, that the short term of managerial contracts was intended as an incentive (cf. Benaissa 2007, 80, 9n.), and such inducements may not have been necessary for the *paramonarioi* (who could have even been *enapographoi*).

171. Cf. *P.Oxy.* I 136.32, LV 3804.154.

172. Cf. Morelli 1996, 51 and n. 24, as well as Rathbone 1991, 91–92.

173. E.g., *P.Oxy.* XIX 2239, a labor contract from the estate of Flavius Ioannes for an overseer (*epikeimenos*). *Opsōnion* is expected, but *misthos* appears, and the *misthos,* moreover, includes payments of wheat, wine, and barley.

174. *P.Oxy.* LVIII 3952.55–56.

175. Note also *P.Stras.* I 40.46, where μισθοῦ ἤτοι διαρίου καθ' [read κατ'] ἔτος appears.

176. τοῖς π̣α̣[ιδαρ(ίοις) Αἰγ]υ̣π̣τ̣ί̣(οις) ὑ(πὲρ) ὀψωνί(ων) θ ἰ̣[νδ(ικτίωνος)] οἴ[ν]ο̣υ̣ κ[νίδ(ια)] ̣τμη.

177. Cf. *P.Oxy.* LVIII 3960, pp. 118–19.

dren" were juridically free.[178] There are several indications that this was their condition. For one, estate *paidaria Aigyptia* ("Egyptian *paidaria*") appear with patronymics in *P.Princ.* II 96, and slaves did not have legal fathers.[179] In the same Princeton papyrus—as in 3960—they are also recipients of salaries, *opsōnia,* which itself suggests free status. Finally, the Princeton text reveals several *meizoteroi* among their number, and we have already seen that these were high-level figures—including, probably, commanders of the *buccellarii* and, whatever their actual role, unlikely candidates for servitude.[180] A military connection for a portion of the estate *paidaria* is indicated, in any case, by *PSI* VIII 953, in which the *paidaria Gothika* ("Gothic *paidaria*") and their families appear with some frequency;[181] doubtless these are the Oxyrhynchite *oikos'* contingent of *buccellarii.*[182] Indeed, the dependency and patronage implied by the very name *paidarion*—a relationship, moreover, that may typically have been permanent[183]—are hardly inconsistent with the *buccellarii,* whose origin may be traced in part to the (free) *domestici* or *oikeioi* of late antiquity.[184]

P.Oxy. XIX 2244 fr. 1 also refers to *paidaria:* its first line should read †̣

178. Sarris (2009, 100) follows Rea and places stock in *eleutheroi.* He assesses the *oikos' paidaria* as follows: "On the directly exploited part of the estate or *autourgia* we encounter bands of workers of an apparently servile condition described as *paidaria,* some of whom are likely to have been engaged in cultivation." This has no basis in the evidence. Banaji (2007, 186) suggests that the *paidaria* were "domestic and administrative staff" and proposes (in n. 107) the intentionally ambiguous rendering "helpers."

179. Cf. also *PSI* VI 709 with Beaucamp 1992, 58 n. 38. The *paidiskē* in this Florentine text no doubt also appears in *PSI* III 194, where her patronymic has been read Hyakinthos (as opposed to 709's Iakinthos).

180. Note also *P.Oxy.* LXXI 4835, 9n., which cautions that "the term [*paidarion*] need not refer to menial or 'domestic' servants but could designate estate employees of all sorts, including ones involved in the administration and management of the estate."

181. The editio princeps has γοθθ, left unresolved; the DDBDP gives the entries as τοῖς παιδαϱ(ίοις) Γόθ(οις), but I suspect the Latin Gothicus. These *paidaria* appear alongside their "*gynaikes* [women/wives] and others" in lines 17, 32, 46–47, and 84–85. When the quantity of wine distributed is extant, it is always 219 *dipla* per month, though in lines 84–85, this figure is reduced by 3 *dipla* for a decease (in l. 85, ἀπὸ δι(πλῶν) σι[θ should be read for ἀπὸ δι(πλῶν) ̣ ̣ [ ]). In line 17, the *paidaria* receive a cash payment in lieu of wine; if 219 *dipla* were adaerated, the tariff is almost exactly 30 *dipla* per *solidus,* a commutation figure attested elsewhere in the text (l. 57). These Gothic *paidaria* could have appeared in the missing part of *P.Oxy.* LVIII 3960, or they might not have been present any longer (because of the Sasanian conquest, e.g.). Or possibly they are included among the *Aigyptia.*

182. The payments of wine to them in 953 are for their maintenance while "in barracks." The distributions in this text that mention *buccellarii* explicitly are either for the support of external troops that have come to Oxyrhynchus (i.e., the *paramenontes*) or for those leaving it on missions; cf. pp. 112, 127 above.

183. E.g., all but one of the *meizoteroi* in *P.Princ.* II 96 are retired.

184. See Schmitt 1994, 154.

β̣[ρέ(ουιον)] ὀψωνίου τῶν παιδαρ(ίων), "salary schedule for the *paidaria*," instead of (ὑπὲρ) [λόγ(ου)] τ̣[α]π̣ι̣δαρ(ίων), read ταπηταρίων.[185] The *opsōnia* in question were in wine, altogether 4,600 *dipla* (l. 2). This heading no doubt stood at the beginning of a lengthy list of names. The steward's account *P.Oxy.* XXVII 2480 happens to include such a list (ll. 126ff.), and its editor has identified the individuals on this schedule as "in the private service of the Apion household."[186] Though it is clear that 2244 fr. 1 does not belong to 2480,[187] the list in the latter text included at least some *paidaria*. Family groups are ubiquitous (cf. the preceding paragraph), and line 293 (on one of the loose fragments) reads, "To the Egyptian *paidaria* and their wives and others inscribed below";[188] this may, in fact, have been the heading of 2480's schedule.[189]

Observations concerning the remuneration of particular kinds of workers on the estate are difficult to make; the samples for those professions that appear more than once in the documentation are small. Among broader groups like the *paidaria*, variation might seem likely, and gender- and age-based distribution patterns are, in fact, detectable in the "household" list of *P.Oxy.* XXVII 2480. If we consider only text that is certain, the mode for men is 36 *dipla* (or 3/month);[190] for women, it is half of this (18 *d*, or 1.5/month). All women identified as wives receive this *opsōnion*, as does a sister. The mode for men explicitly identified as "sons" (i.e., with υἱός; ignoring filiations indicated simply by the genitive of the father's name) is also 36, though these individuals average somewhat less than men as a whole (28.67, $N = 10$, compared to 31.25, $N = 24$). Two distributions of 12 *dipla* among the sons, as well as one among the daughters, were probably for children. Most figures, as might be expected for an annual *opsōnion*, are cleanly divisible by twelve.

Even the largest of these *opsōnion* figures (i.e., the 36 *d* for men) is clearly a consumption allowance. It amounts to 157 liters annually,[191] a volume falling

185. For the reading † β̣[ρέ(ουιον)], see Gonis 2004a, 200–201. Gonis also corrects τ̣[α]π̣ι̣δαρ(ίων), but note already Hickey 2001, 171, where it was likewise suggested that line 3's ἀπὸ μειζ(όνων) should be read ἀπὸ μειζ(οτέρων).

186. *P.Oxy.* XXVII 2480, p. 183.

187. The possibility of a connection is broached in 2480, 167n., but see already Rea 1963, 181: "[T]he fragment [of 2244] does not appear to belong to 2480 because of slight differences in the hands and in the qualities of the papyrus on which they are written."

188. τοῖς ἐγγεγραμμ[(ένοις) παιδαρ(ίοις)] Αἰγυπτίοις καὶ [γ]υ[ναιξὶ καὶ ἄλλοις (*BL* X 149). A recent photograph suggests that the papyrus has deteriorated significantly since publication, but I read ἑ̣ξῆς after τοῖς.

189. The first two words are in ekthesis.

190. This is the same amount by which the monthly distribution to the *paidaria Gothika* is reduced when one of their number dies; see n. 181.

191. As already noted, the standard *diploun* in *P.Oxy.* XXVII 2480 held 8 *sextarii.*

between Clarysse and Vandorpe's estimate of wine consumption in the Ptolemaic Fayum (100 l/person/year) and Tchernia's figure for Rome (.8–1.0 l/adult male/day)[192] and in closest accord with the datum provided by Purcell for prerevolutionary Paris (ca. 1780; 120 l.).[193] The 4,600 *dipla* in *P.Oxy.* XIX 2244 fr. 1 would have furnished annual rations for 127 adult male *paidaria,* while the 219 *dipla* distributed monthly to the *paidaria Gothika* in *PSI* VIII 953 correspond to 73 adult men. In both cases, the total number of *paidaria* would have been higher because women and children were among the recipients.[194] The certain figures from the "household" list in *P.Oxy.* XXVII 2480 yield a mean of 28.1 *dipla* ($N = 31$), which suggests approximately 164 and 94 *paidaria,* respectively. If estate housing was one of the defining characteristics of the *paidaria,* the typical individual of this status may well have received little more than a subsistence allowance (the bulk of which would have been given in kind).[195]

A maximum of 16,065 *knidia* is available for permanent employees and their dependents in *P.Oxy.* LVIII 3960.[196] Division of this figure by the mean distribution in 2480 (i.e., 28.1 *d*) yields a permanent workforce (inclusive of family members eligible for wine) of roughly 572 individuals.[197] This estimate is almost certainly too high. Already based on a maximum, it would need to be reduced, for example, if the *paramonarioi* received more than a consumption allowance, and in any case, important and senior employees surely received more than 36 *dipla* annually. I address the significance of this estimate in the next chapter.

Payments of wine to estate managers are attested by *P.Oxy.* XVI 2051r. The second column of this papyrus, which probably preserves scribal practice, has a series of entries (six in all) like the following:[198]

---

192. Fayum: Clarysse and Vandorpe 1997, 71, where beer is plausibly assumed to be the beverage of choice for the indigenous population. By late antiquity, beer has all but disappeared from the sources; cf. Bagnall 1993a, 32. Rome: Tchernia 1986, 10.

193. Purcell 1985, 15 n. 71.

194. The equal distribution of wine to women in table 4.9.6 (if, in fact, it occurred) is presumably attributable to their presence on a mission (when higher rations were distributed).

195. Cf. Rathbone 1991, 106–16, and note as well *P.Stras.* I 40, in which Aurelius Kollouthos agrees to work as a φαμιλιάριος ἑδραῖος κατάδουλος παῖς for four years at subsistence-level compensation (including 24 *knid.* of wine; the only money is a clothing allowance). Morelli (1996, 49) sees *paramonē* in this agreement, but note *P.Heid.* V, p. 293.

196. This is the remainder when all nonstaff distributions are subtracted from the steward's *hypodochē.*

197. Again, the *knidia* of 3960 are almost surely the same size as the *dipla* in 2480; see n. 35 above.

198. Scribal practice: See p. 44 n. 34 above. I present the sample entry in column form to ease reading; in the papyrus, the entries are linear.

σμήμ(ατος) καὶ ἐπιτρυγ( ) οἴνου δι(πλᾶ) κ, for *smēma* and *epitryg( )*, 20 *dipla*[199]
ἀντιγεούχ(ῳ) οἴνου δι(πλᾶ) δ, to the landlord's representative, 4 *dipla*
διοικ(ητῇ) δι(πλᾶ) ς, to the *dioikētēs*, 6 *dipla*
χαρτ(ουλαρίῳ) δι(πλᾶ) β, to the secretary (*chartularius*), 2 *dipla*
ἐπίκ(ειμένῳ) δι(πλᾶ) ε,[200] to the overseer, 5 *dipla*
προ(νοητῇ) δι(πλᾶ) δ, to the steward (*pronoētēs*), 4 *dipla*
ἀρχισυμμ(ά)χ(ῳ) δι(πλᾶ) β, to the chief messenger, 2 *dipla*
σταλάγμα(τος) ἐπιτρυγ( ) δι(πλᾶ) ε, for *stalagma epitryg( )*, 5 *dipla*[201]

The volume of wine associated with σμήματος καὶ ἐπιτρυγ( ) is variable (20, 30, or 40 *dipla*), but the other quantities remain constant. The payments to employees might be interpreted as *synētheiai* attached to individual leases,[202] but their absence from Text I suggests a charge (or possibly a bonus from the estate) on the level of at least the *epoikion*.[203] The schedule may allot more *dipla* to overseers (if local foremen are meant) than to (some) managers because the latter received wine from more sources (the *pronoētēs* from the holdings of his *prostasia*, the *dioikētēs* from the settlements in his *dioikēsis*, and the *antigeouchos* from the entire estate).[204] Of these managers, only the *pronoētai* have labor contracts preserved in the extant estate documenta-

199. I would suggest rendering *smēma* (or *smēgma;* Gignac, *Grammar* 1:177) as "grease." *Smē(g)ma* need not be obtained from grapes; cf., e.g., *P.Ryl.* II 230 (vetch) and *SB* XXIV 16285 (ἀπὸ κικίνου, castor seeds). For its use in a vineyard, presumably as a lubricant for a press and/or irrigation machinery, see *P.Herm.* 38.6. Note also *P.Ryl.Copt.* 159, where the Egyptian equivalent ⲁⲛϫⲓⲣ appears among the expenses to be paid by the tenants. The description of *P.Oxy.* XVI 2051r renders ἐπιτρυγ( ) as ἐπιτρυγ(ῆς), a hapax. For a discussion of alternatives (including division into two words, i.e., ἐπὶ τρυγ( )), see Hickey 2001, 95–96.

200. *BL* IX 192.

201. For *stalagma,* see Chouliara-Raïos and Wagner 1990, 72–73, where it is proposed that it might be the equivalent of *stagma,* which, in turn, is described as "a sweet drink made from a mixture of at least two or three of the following ingredients: must, i.e., sweet new wine; wine; honey" (73–74); cf. also *P.Kellis* IV 96, p. 46. Chouliara-Raïos and Wagner do not seem to know *P.Lond.Copt.* 1041.2, which has "two thousand *kollatha* [vessels] of cooked [or "boiled"] *stagma,*" ⲛⲥⲛⲁⲩ ϣⲟ ⲕⲟⲗⲁⲑⲟⲛ ⲥⲧⲁⲕⲙⲁ (read ⲥⲧⲁⲅⲙⲁ) ⲉⲩ[ⲡ]ⲟⲥⲉ—a description that is suggestive of Medieval and Modern Greek ἀπόσταγμα (cf. their p. 72, where this is described as "a liqueur or oil obtained by distillation [eau de vie, rose oil])."

202. Cf. p. 108 above.

203. Cf. *P.Oxy.* LV 3805.34 (*synētheiai).*

204. For the *epikeimenos,* see *P.Oxy.* LV 3805, 35n. There were multiple *archisymmachoi;* cf. *SB* XX 14077. The *chartularii* are a more complicated matter (cf. Hardy 1931, 94–95); perhaps their distributions here were tied to their supervision of the vintage (cf. p. 68 n. 42 above).

tion.[205] Both examples are vague about remuneration—the better preserved simply states that the contractor will accept the same salary as his predecessor[206]—but an *opsōnion* of 24 *artabai* of wheat and 2 *solidi* less 5 *keratia* is attested in their accounts.[207] The discrepancy between this and the "entry fee" for the position has long been noted;[208] other scholars have suggested additional sources of income,[209] but wine has never been considered in the equation.[210]

Clear instances of *estate* distribution beyond consumption level do appear in the dossier, and unsurprisingly, none of the classifiable recipients are "mere" laborers. *P.Oxy.* XXVII 2480.63, for instance, records a 75-*dipla opsōnion* for a *scholastikos* ("advocate"), and a notary seems to receive a vintage bonus (*hyper trygētikōn*) of 120 *dipla* (60 *d* × 2 indictions) in *PSI* VIII 953.66–67.[211] Sergios, the son of Phoibammon (a retired *meizoteros* or *apo meizoterōn*), is another recipient of a vintage bonus in this account, but his was in cash;[212] at 4 *solidi* less 16 *keratia* (or 4 *sol.* on the private or "estate" standard)—if commuted at the rate of 30 *dipla* per *solidus* attested elsewhere in the text (l. 57)—it was the equivalent of 120 *dipla*. *Meizoteroi* themselves, both active and retired, are also attested as recipients of greater quantities of wine.[213] In *PSI* III 191, for example, a certain Serenos *meizoteros* receives 50 *dipla* of vintage (*palaios*) wine by order of Apion II, while an *apo meizoterōn* is the recipient of 100 *dipla* in

205. Certain texts: *P.Oxy.* I 136, LVIII 3952.

206. *P.Oxy.* I 136.31–32, δέξασθαί με τὸ ἐμὸν ὀψώνιον κατὰ μίμησιν τοῦ πρὸ ἐμοῦ προνοητοῦ.

207. See *P.Oxy.* LV 3804, 154n.

208. The entry fee is called *paramythia* in *P.Oxy.* I 136, *eisbatikon* in *P.Oxy.* XIX 2239 (a contract for an *epikeimenos* on another estate). The *paramythia* is 12 *solidi* on the Alexandrian standard in 136; in *P.Oxy.* LVIII 3952, it includes hayseed in addition to a sum of *solidi* (now lost). The *pronoētēs* also had to furnish a premium equal to 15 percent of the *sitos* due from his *prostasia*, but this was extracted from the *geōrgoi*.

209. Cf., e.g., Hardy 1931, 92–93; Johnson and West 1949, 60–61; and Maresch 1994, 130–31.

210. *BGU* I 310 (= *SB* I 5330), an Arsinoite *pronoētēs* contract from 663(?), mentions customary *synētheiai* and *trygētika* (l. 15), but they are not enumerated. Wine is part of the steward's *opsōnion* in BL inv. 2219 (Oxyrhynchus [not connected to Apiones], 496; forthcoming from Keenan and myself).

211. Azzarello, *P.Köln* X, p. 206, n. 3, suggests that these *dipla* are to be distributed to workers by the *notarios*, an interpretation that is supported by *P.Oxy.* XVI 2032.63 (discussed below). *P.Oxy.* XXVII 2480.151 likely concerns a *notarios;* only]ταρ( ) remains, but *notarioi* appear elsewhere in this section of the text (ll. 262–65, 268–70). Line 151 alone preserves the amount distributed (36 *dipla*), but this part of the account concerns *opsōnia*, and supplemental distributions (like *trygētika*) would have been recorded elsewhere.

212. *PSI* VIII 953.19–20. Sergios' own title is damaged and undecipherable. Though the position of *meizoteros* might stay within a family (cf., e.g., *P.Princ.* II 96.4, 8), the word cannot be read here. Since the *meizoteros* could be a military commander, perhaps Sergios was one of the *buccellarii* (cf. *P.Oxy.* XVI 1903.8). He may reappear as a recipient of wine in *PSI* VIII 956.23–24.

213. Again, the 36-*dipla* distribution to the *apo meizoterōn* in *P.Oxy.* XIX 2244 fr. 1.3 is an *opsōnion;* additional compensation would have appeared elsewhere.

PSI VIII 953.53. The *oinocheiristēs* himself received a share of the wine in his charge; probably tagged *hyper rhyseōn* ("for the vintage"),[214] in 953.3 it amounts to 73.5 *dipla.* This is almost exactly .2 percent of the total *analōmata* in the immediately preceding line, but the sum distributed may, in fact, have been fixed.[215] A certain Philoxenos, occupation unknown, has been allotted 72 *dipla* as his *opsōnion* in 953.24.[216]

Evidence in the dossier for the payment of *misthoi* ("wages") in wine is sparse; the only explicit instance is *PSI* VIII 953.3–4, where a *kompastēs* (the individual testing the soundness of wine jars) receives a wage of 5 *dipla.* As *misthos* implies,[217] this was not a year-round post. Other casual workers and specialists—like Menas from the (Hermopolite?) settlement Thalmaon(?), who is variously described as a boatman and a shipwright (*nautēs/naupēgētēs*) and was working on the great estate *karis*[218]—had at least a portion of their wages commuted into wine and other foodstuffs. Menas received 4 *dipla* per month for three months, plus, perhaps, a bonus at his job's end; the total value of the distribution was about half of a *solidus.*[219] Menas may also have been given some wheat, meat, and oil. If the wine that he received was for him alone (i.e., if the payments did not include remuneration for any helpers),[220] his monthly distribution was 33 percent higher than the monthly consumption allowance of male *paidaria* (cf. above). The distributions to Menas are all tagged with the rather generic *logō analōmatōn.* As Morelli has suggested in regard to an earlier contract involving visiting *naupēgoi* from Ptolemaïs (*P.Oxy.* XVI 1893, 535), payment in kind was probably convenient for individuals like Menas who were working away from home.[221] This observation may also be relevant for the *laotomoi* (stonecutters) from the *epoikion* Philostratou who received 6 *dipla* of wine in connection with their work on the pillars for the new Apion *proastion* (suburban estate house); the quality of the wine (*palaios,* "vintage") could indicate a bonus.[222]

214. The text only has ὑπὲρ ῥρ( ); the editor's suggestion, *roga* ("ration"), seems less appropriate than *rhysis.*

215. It is described as κατὰ μίμησιν ιε ἰνδικτίονος, "in conformity with the 15th indiction," but this might refer to its calculation.

216. His brother and sister appear in the following line.

217. But see above, p. 130.

218. For *karis,* see *PSI Com.* 20, 4n.

219. *P.Oxy.* XXVII 2480.1, 15, 26–27; possible bonus: l. 36.

220. Note that *P.Oxy.* XXVII 2480.33–34 reads, τοῖς ναυπηγήταις καὶ καλαφάταις καὶ τέκτοσιν καὶ ἄλλοις λόγῳ φιλοτιμίας ἐν τῇ ἡμέρᾳ ἐν ᾗ ἐβλήθη ἡ μεγάλη καρὶς εἰς τὸν ποταμὸν, "To the shipwrights and caulkers and carpenters and others on account of largesse on the day on which the large *karis* was launched into the river." It is unclear if Menas is also included among these *naupēgētai*; elsewhere he appears alone, but no other shipwrights are named in the extant portion of the account.

221. Morelli 1996, 48–49.

222. *PSI* III 193.

The Hippodrome

Additional distributions of wine tied to special events do not have a separate entry in the extant portion of *P.Oxy.* LVIII 3960, but they have certainly been mentioned in the preceding sections: Brumalia, Christian feast days, the new year, birthdays of the *geouchoi,*[223] and the vintage are some of the occasions that have figured in the discussion. A related (and well-documented) set of distributions are those connected to the hippodrome. In their case, however, the fiscal element is manifest.[224] Between the fourth century and the sixth, the *hippotrophia,* "the maintenance of stables and of the personnel of the colors [i.e., the 'circus factions'], the organization of spectacles and festivals,"[225] was transformed. Financial support of the circus, once a liturgy for the town councilors (*curiales*), became a city service; the hippodrome was now the beneficiary of municipal taxes jointly administered (under the system of fiscal shares) by the aristocratic *oikoi.*[226] In the wake of this transformation, the Blue and Green *merē,* or factions, functioned as "conventional subdivisions of a common institution, between which the authorities distributed receipts [income], responsibilities, and tasks . . . they came to be perceived only as parts of a whole."[227] This renders comprehensible the glorious house's support of both *merē* during a period when its master was apparently a partisan of the Blues.[228] Nevertheless, table 4.10 reveals that none of the wine distributions attested in estate documentation are explicitly connected to the Greens. Most remarkable are the veterinary expenditures (4.10.1–13), which dominate the list.[229] The other entries require little additional comment. One might have expected more distributions to charioteers and representation of other figures, for example, the *hippodiōktai* (chasers), *hippokomoi* (grooms), and *aphetai*

223. Note also *PSI* VIII 953.80, 89 (the birthday of Apion II).

224. This is not controversial; cf. both Gascou (e.g., 1976b) and Liebeschuetz (e.g., 2001, 180), who have divergent views about civic finance in this period. For a possible fiscal element in other spectacles, cf. pp. 108–9 above.

225. Gascou 1976b, 192.

226. So Gascou 1976b; Liebeschuetz (2001, 206), in contrast, sees the involvement of landowners like the Apiones as a continuation of the "traditional munificence of decurions and other magnates." For extensive documentation concerning the *hippotrophia* (racing liturgy) in the fourth century, see *O.Ashm.Shelton* 83–190; Shelton 1990; Gonis 2002a.

227. Gascou 1976b, 199–200.

228. Cf. Gascou 1976b, 195 n. 3.

229. For discussion of certain points, see the notes attached to table 4.10. Regarding the use of wine in equine care, it may also be noted that a *hippiatros* is the buyer-creditor in a sale of wine in advance (*P.Oxy.* LXI 4132), while another horse veterinarian receives compensation for wine turned to vinegar in *P.Oxy.* XVI 1974 (the original transaction was another sale in advance). *Hippiatroi* would have been on hand for races (cf. *O.Ashm.Shelton* 83, 131, 144, and p. 77), and the fragmentary *PSI* VIII 955.14–15 may have contained a distribution connected to such work.

TABLE 4.10. Distributions Connected to the Hippodrome

Text	Steward/Date	Recipient(s)	Amount	Notes
1. *P.Oxy.* XXVII 2480.22–23	Phib/Thoth (certainly on the 9th)	horses τοῦ ἱππικοῦ (sc. μέρους Βενέτων, as elsewhere?)	2 *dipla* + 2 *d* = 4 *d*	—2 διπλᾶ εἰς βροχήν, *thea* on Thoth 9[a] —2 διπλᾶ εἰς βροχήν, θέα τοῦ Στεφα-νίωνος (in Thoth)[b]
2. *P.Oxy.* XXVII 2480.10–11	Phib/Thoth (certainly on the 20th)	horses τοῦ ἱππικοῦ μέρους Βενέτων	1 *diploun* + 1 *d* = 2 *dipla*	—1 διπλοῦν εἰς βροχήν, θέα τῶν Στεφανίων (in Thoth) —1 *d* εἰς βροχήν, *thea* on Thoth 20
3. *PSI* VIII 953.77	Serenos/Thoth–Mecheir 4	horses [ed. pr.: τῶν ἱππ( ); cf. *BL* V 125] τοῦ ἱππικοῦ μέρους Βενέτων	1_ *dipla* (clearly not same rate and/or quantity of horses as in 4.10.4)	—*eis brochēn* (ed. pr.: [– – –]οχην; cf. *BL* V 125)
4. *PSI* VIII 953.42	Serenos/ Pharmouthi 11–Pachon 2	horses [ed. pr.: τῶν ἱππ( ); cf. *BL* V 125] τοῦ ἱππικοῦ μέρους Βενέτων	16 *dipla,* 22 days = .72 *d*/day	—*eis brochēn* (ed. pr.: εἰ[ς χρεί(αν)]; cf. *BL* V 125)
5. *P.Oxy.* XXVII 2480.81	Phib/Pharmouthi 18	horses τοῦ ἱππικοῦ	3 *dipla*	—εἰς βροχήν, θέα τῆς Κράστης
6. *P.Oxy.* XXVII 2480.83	Phib/Pharmouthi 20	horses τοῦ ἱππικοῦ μέρους Βενέτων	1 *diploun*	—εἰς βροχήν, θέα τῆς Κράστης
7. *P.Oxy.* XXVII 2480.90	Phib/Pharmouthi 20	horses τοῦ ἱππικοῦ μέρους Βενέτων	2 *dipla*	—εἰς βροχήν, θέα τῆς Κράστης
8. *P.Oxy.* XXVII 2480.99	Phib/Pachon 21	horses τοῦ ἱππικοῦ μέρους Βενέτων	1 *diploun* wine, 1 *d* vinegar	—εἰς βροχήν
9. *P.Oxy.* XXVII 2480.108	Phib/Epeiph 7	horses τοῦ ἱππικοῦ μέρους Βενέτων	1 *diploun*	—εἰς βροχήν

TABLE 4.10—*(Continued)*

Text	Steward/Date	Recipient(s)	Amount	Notes
10. *P.Oxy.* XXVII 2480.118	Phib/Epeiph 15	horses τοῦ ἱππικοῦ μέρους Βενέτων	3 *dipla*	—εἰς βροχήν
11. *P.Oxy.* XXVII 2480.97	Phib/Pharmouthi 14	horses τοῦ ἱππικοῦ μέρους Βενέτων	1 *diploun* vinegar	—εἰς ἐπιμέλειαν[c]
12. *P.Oxy.* XXVII 2480.98	Phib/Pharmouthi 26	horses τοῦ ἱππικοῦ μέρους Βενέτων	1 *diploun*	—εἰς ἐπιμέλειαν
13. *P.Oxy.* XXVII 2480.101	Phib/Pachon 4	horses τοῦ ἱππικοῦ μέρους Βενέτων	2 dipla	—εἰς ἐνχυματισμόν[d]
14. *P.Oxy.* XXVII 2480.82	Phib/Pharmouthi (θέα τῆς Κράστης)	charioteers and *philētai*[e] τοῦ μέρους τῶν Βενέτων	6 *dipla*	— λόγῳ ἀναλώματος, θέα τῆς Κράστης
15. *P.Oxy.* XXVII 2480.107	Phib/Epeiph (θέα τοῦ Πηχισμοῦ)	charioteers *merous Benetōn*	3 *dipla*	—λόγῳ ἀναλώματος, θέα τοῦ Πηχισμοῦ[f]
16. *P.Oxy.* XXVII 2480.28	Phib/Phaophi	*chalkeis*	4 *dipla*	—making keys τοῦ ἀγωγοῦ τοῦ στάβλου τῶν ἵππων τοῦ ἱππικοῦ μέρους Βενέτων
17. *P.Oxy.* XXVII 2480.105–6	Phib/Pauni	δεσπόται τῶν τιρώνων ἵππων	4 *dipla*	—*despotai* described as selling τὰ αὐτὰ ἱππάρια εἰς λόγον τοῦ ἱππικοῦ; λόγῳ ἀναλώματος
18. *PSI* VIII 953.62 (567/68)	Serenos / ?	*thyrouros* of the landlord's box at the circus	12 *dipla*	

[a]Cf. Perpillou-Thomas 1993a, 255, where Thoth celebrations in the late period are linked to pre-Christian festivals associated with the beginning of the Egyptian year.

εἰς βροχήν: Although the editor of *P.Oxy.* XXVII 2480 remarks (10n.) that wine is mentioned as a drink for horses in the *Iliad,* and despite the fact that numerous oral uses of wine and vinegar are attested in equine medicine (cf., e.g., Col., *De re rustica* VI.30, 34), it is probable—all the more so given the references in this account to distributions *eis enchymatismon* (cf. n. d below)—that some sort of topical application must be

(continues)

TABLE 4.10—*(Notes continued)*

meant, cf., e.g., *Hippiatr. Berol.* XXX.6 (Oder and Hoppe, p. 152), χρῶ δὲ καὶ ἐμβροχῇ οἰνελαίῳ καὶ ἀφρονίτροις, καὶ ῥάκιον ἐπιτίθει κατὰ τοῦ νώτου μετὰ ἐμβροχῆς θερμῆς. The editor of 2480 also notes, "[T]he possibility that this wine was used as a massage cannot in this instance be excluded."

In most cases, the tag *eis brochēn* is explicitly associated with a *thea*—in other words, with racing—and it is possible to associate the other entries with the expression to such activity. The chronological span in table 4.10.4 would seem to include the θέα τῆς Κράστης, and both 4.10.11 and 12 could have some connection to this event as well. Table 4.10.9 and 10 might be connected to the θέα τοῦ Πηχισμοῦ (cf. 4.10.15). Table 4.10.3 might concern races held in Mecheir; cf. *P.Oxy.* I 152 (precise date of races not specified), as well as *O.Ashm.Shelton,* p. 74 (unlike here, at the end of the month). The θέα τῆς Κράστης, incidentally, could be behind the Pharmouthi races in *O.Ashm.Shelton.* The editor of 2480 (64n.) suggests that this *thea* was connected to the harvest and hypothesizes a link with γράστις/κράστις. Though there could be an agricultural connection, it is hard to ignore the fact that this spectacle falls shortly after Easter. The earliest secure date connected to the *thea* is Pharmouthi 18 (4.10.5); this happens to be the Octave of Easter (if my date of 580/81 is accepted for 2480), and it may be noteworthy that the "golden races," the first and most solemn races of the Constantinopolitan season, took place on the day after the Octave (Const. Porphyr., *Cer.* I 77 [= Dagron et al. 2000, 11–19]). Of course preparations would have begun days before the races themselves (cf. also, e.g., 4.6.25, 26).

[b]This *thea* remains opaque. The ed. translates "the spectacle of the crowns." Crowns suggest martyrs, and there was a commemoration of St. Stephen on Thoth 15 (cf. *P.MorganLib.* 51, p. 70). There also was an Apion *epoikion* Stephanionos, cf. *P.Oxy.* XIX 2244 (and XVI 2034). A connection with a *dies imperii* is improbable.

[c]εἰς ἐπιμέλειαν ("for care") needs no explanation, but it is too generic to allow for precise commentary on the use of the wine. The tag is also used to describe some of the *oikos*' expenditures on the animals of the *cursus velox;* cf. *PSI* VIII 956.28.

[d]The editors of the papyrus comment, 100n.: "drink, massage?" "Injection, instillation," the definition given by *LSJ,* is unsatisfying. "Infusion," which is of course related to the Latin *infundere* (the verb employed by Columella for the administration of various equine remedies by pouring down the throat, cf. VI.30.3, 9), seems better; cf. also Sophocles' translation of ἐγχυματίζω ("to infuse, as a medicinal fluid") and now McCabe 2007, 116–17 (to administer a drench or potion). A cursory examination of the *Hippiatr. Berol.* reveals wine-based mixtures infused through the right nostril (Oder and Hoppe, p. 161, XXXII.3, e.g.) and the mouth (pp. 246–47, LVI.2, e.g.). See also p. 386, where various recipes for *enchymatismoi* appear; those described as "simple," *thermantikos* ("warming," presumably), and *epi chloēs* (at the arrival of spring) contain wine.

[e]See Gascou 2008, 347.

[f]Cf. Perpillou-Thomas 1993a, 116, who writes that this *thea* is "probably tied to the inundation of the Nile; it took place on the 16th and perhaps up to the 18th of Epeiph (10–12 July Julian), also the date on which the Nile overflowed its bed; the noun πηχισμός, 'measurement by means of the cubit,' reminds us that the height of the waters was measured in this way on the Nilometer."

(starters).[230] Table 4.10.17 remains opaque: it seems that the proceeds of the sale of these *hipparia* (ponies) went to the circus, but the identity of the sellers is a mystery. Table 4.10.18 is probably a private Apion expense. A *hippotrophos* named Anastasios twice appears in one of the fragmentary *oinocheiristēs* accounts (*PSI* VIII 955.1, 32),[231] but this individual was presumably an estate

230. For the *hippodiōktai,* as well as other individuals involved in racing, see *O.Ashm.Shelton,* pp. 74–77.

231. Here it may be noted that *SB* XX 15181 fr. 6 joins to the left side of 955.33–37.

employee; at any rate, he clearly was no longer responsible for the expenditures of the circus.[232]

Outflows, Part 2: Sales of Wine

As we have seen, the data concerning estate expenditures in wine are extraordinarily extensive due to the survival of three well-preserved wine stewards' accounts. The existence of these rich texts, when considered alongside the known involvement of *oinocheiristai* in the wine trade,[233] might lead us to expect references to sales in their accounts. None appear—in fact, the silence in the entire dossier is deafening. The only extant document exclusively concerning a sale of estate wine is *P.Oxy.* VIII 1131.[234] There, the *nomikarios* Ioulianos promises to pay the *oikos' oinocheiristēs* 2 *solidi* less 8 *keratia* (i.e., 2 *sol.* on the private standard) for some wine that he has already received; 1131 is thus a sale on credit, or *Kaufpreisstundung*.[235] Since payment will occur "when the glorious house takes stock of its arrears,"[236] it seems more likely that the seller is, in fact, the *oikos* and not the *oinocheiristēs* (selling his own wine). As is typical in texts of this type, the quantity of wine purchased is not specified,[237] and illegal (i.e., usurious) interest naturally comes to mind.[238] In a general discussion of *Kaufpreisstundungen*, Jördens does not discount interest but cites other factors: "Although we scarcely will want to reckon with interest-free loans granted out of pure generosity, the limited terms [of the agreements] and the small number of the examples could still very probably indicate a contract resulting from accidental and temporary liquidity difficulties, especially since nearly half of the cases concern the basic foodstuff wine."[239] In the case of 1131, Ioulianos' position as *nomikarios* may be significant. The somewhat obscure *nomikarioi* had some connection with both tax collection and official transport,[240] and the

232. Cf. Gascou 1976b, 192–93 n. 3.

233. Cf. table 4.2 and *P.Oxy.* XVI 2032 (discussed imminently). Other potential wine sellers (*oinopratai, oinopōlai, oinemporoi*, etc.) make but one (probable) appearance in the dossier (*SB* XVI 12608; see pp. 128–29 above), and wine belonging to estate stocks does not seem to be at issue in this case.

234. The text is dated to the fifth century in the editio princeps, but it actually belongs to the mid-sixth; cf. *BL* X 141. A much earlier text, *P.Oxy.* LXIII 4391 (471), might also concern a sale (cf. its 2n.).

235. See Jördens 1993.

236. ἐν τῇ λοιπογραφίᾳ τοῦ ἐνδόξου οἴκου; presumably this occurred at the end of the fiscal year. *P.Oxy.* XVI 2032 was probably employed during this process; *P.Oxy.* XVI 1918r is its result.

237. Cf. Jördens 1993, 272.

238. Cf. the arguments in Bagnall 1977 about advanced sales (*Lieferungskäufe*).

239. Jördens 1993, 280.

240. Cf. *P.Oxy.* LV 3788, 2n., LVIII 3942, 9–10n., LIX 3985.

Apiones were certainly involved with this one's work (cf. below); he may have even been under contract to the *oikos*.[241] Such relationships could have entitled Ioulianos to more favorable treatment; the same might be said if the transaction involved wine destined for state purposes.[242]

Ioulianos reappears in *P.Oxy.* XVI 2032 (540–41 according to *BL* VI 105).[243] The first column of this long account is headed † ἔχθ̣(εσις) χρυ̣[σικῶν, and there are many cancelled (crossed-out) entries.[244] *Echthesis* can simply mean "schedule" or the like, but it often heads a list of arrears. In this case, there is no ambiguity: *P.Oxy.* XVI 1918r cols. i–ii, which summarize the entries in 2032, make it clear that these are liabilities against the estate's net cash income (see 1918r.17; cf. v.13).[245] Given Ioulianos' promised payment at the *loipographia*, some trace of 1131 might reasonably be sought in 2032, but it does not appear; the only agricultural commodity with which Ioulianos is explicitly involved is grain (l. 16). Wine does figure, however, in three entries in the account. *P.Oxy.* XVI 2032.62 reads, "Left with [or 'in the hands of'] Pamouthis the wine steward for the price of wine, [...] ½ 1/8 *solidi* on the private standard, less 604 [*keratia*], with the 4 1/2 *solidi* less 21 [*keratia*] of/for Ision the *exceptor*";[246] the first sum has been corrected from 159 ½ *solidi* less 697 ¼ *keratia*.[247] Two interpretations of this entry present themselves most immediately: either Pamouthis has been advanced money to purchase wine for the estate, a transaction that he had not consummated at the time of the *loipographia;* or Pamouthis, like Ioulianos, has been sold a quantity of wine under a *Kaufpreisstundung*, and

241. *P.Oxy.* XVI 2024.10–11 mentions *nomikarioi* who are also *paktarioi tou oxeōs dromou* (contractors of the *cursus velox*). *Tou endoxou oikou*, "of the glorious house," should probably be understood following *dromou;* cf. *P.Oxy.* I 138.9 (and note *P.Col.* VIII 246.8). Also note *P.Oxy.* LIX 3985 (473), a document in which the village of Takona nominates a *nomikarios* whose duties include collection of village taxes; in 550 (*P.Oxy.* I 133), Takona was under the Apiones' pagarchic authority, but this does not mean that it was still nominating *nomikaroi.* (In any case, *P.Oxy.* XVI 2032.26 no doubt reflects an agreement like 3985.)

242. Mazza (2001, 146) suggests a connection to the *annona militaris* for 1131.

243. Lines 16, 25–26. Also note line 9.

244. Regrettably, the editors do not indicate the changes of hand. The text, one of the few from the highest levels of estate administration, merits a reedition.

245. See Hickey 2008 for further discussion of 1918.

246. π(αρὰ) Παμουθίῳ οἰνοχειριστ(ῇ) ὑπὲρ τιμ(ῆς) οἴν(ου) *vac.* ἰδ(ιωτικῷ) νο(μίσματα) [...]Ḷη´ [π(αρὰ)] χδ σὺν τῶν̣ [read τοῖς] νο(μίσμασι) δL π(αρὰ) κα Ἰσίωνος ἐξέκπτορ(ος) [read ἐξκέπτορος]. For the amounts of money, I follow the editio princeps. In the string Lη´ [π(αρὰ)] χδ, only χδ may now be read with any confidence. (I have removed the editors' underdots, which are unnecessary.) For *para* with the dative as "left with" or "in the hands of," see *P.Oxy.* XVI 2029; *WB* renders "bei." *Exceptor* is often translated as "shorthand writer" (*tachygraphos*); for more nuance, see *P.Oxy* LVIII 3932, 16n. and refs.

247. ρνθ L π(αρὰ) χϙζd´. Note that the figure χϙζd´ has not been crossed out.

he had not yet settled accounts.[248] The former scenario represents an activity already attested for the *oinocheiristai* (cf. table 4.2) and leaves room for an explanation for the curious mention of Ision: perhaps Pamouthis had also been given money to pay the (adaerated?) perquisites (*synētheiai*) for Ision's bureau.[249] The latter proposal, besides echoing *P.Oxy.* VIII 1131, suggests other known behaviors: the risk-averse landlord who sells his surplus (or a portion of it) to an "entrepreneurial" middleman who makes frequent use of credit.[250] Under this scenario, however, Ision's presence remains puzzling (Pamouthis' partner?), while the sum of money involved *greatly* exceeds any attested in the extant *Kaufpreisstundungen;*[251] on the assumption that the ratio of deducted *keratia* to *solidi* in the corrected sum is an accurate guide, Pamouthis would have owed approximately 138 *solidi* less 604 *keratia.* His liability, in any case, would have been a substantial sum, many times more *solidi* than were owed by Ioulianos.

Pamouthis reappears two lines later, this time connected to a much smaller sum (13 *sol.* less 53.5 *ker.* on the "estate" [private] standard; corrected from 31 less 134.5) for the price of vinegar. This second entry is amenable to both interpretations presented above,[252] but another item in 2032 mentioning the price of wine raises doubts that such entries are to be construed as sales on credit. The line in question (63) reads, "Left with [or "in the hands of"] Apion the notary for the price of wine of/for the navvies, 36 *solidi* on the private standard less 144 [*ker.*]."[253] If this entry references a sale on credit to Apion, the mention of the *potamitai* lacks explanation;[254] clearly Apion has been instructed to buy wine to distribute to the *potamitai* on behalf of the *oikos.*[255] It may be, however, that the preposition used in the items under consideration is significant: Pamouthis' entries have *hyper timēs,* not *eis timēn* as in Apion's entry, and the

248. Rather forced seems a third interpretation under which the liability belongs to those to whom Pamouthis has already sold wine; we would expect this party to be mentioned in the entry. Just possibly Pamouthis had sold wine and, for some reason, had not yet delivered the proceeds at the time of the reckoning of arrears.

249. Cf. pp. 106–8 above.

250. For the aristocrat, cf., e.g., Kehoe 1997, 211–12. For the middleman, cf. *P.Worp,* pp. 349–50 and refs.

251. Cf. Jördens 1993, 272–73.

252. Once again, however, the amount at issue is greater than all of those appearing in Jördens' table of *Kaufpreisstundungen;* see the preceding note.

253. π(αρὰ) Ἀπίωνι νοταρ(ίῳ) εἰς τιμ(ὴν) οἴν(ου) τῶν ποταμιτ(ῶν) ἰδ(ιωτικῷ) νο(μίσματα) λϛ [ϛ corr. from L?] παρὰ ρμδ. I depart from the editio princeps at line's end. It has νο(μίσματα) λ παρὰ ροδ [ο corr. from ι?].

254. For these workers and their remuneration, see Morelli 1995. They are most unlikely to have been engaged in the wine trade.

255. Note that Apion has received *solidi* ὑπὲρ τῶν χωμάτων, "for the embankments," in 2032.65.

former collocation appears in sales on credit.[256] Yet, caution seems in order because 2032's scribes sometimes omit the preposition altogether.[257] In any case, the quantity of wine that Apion was charged with acquiring was significant: at the tariff in *PSI* VIII 953.57 (240 *sext./sol.*), it amounts to 8,640 *sextarii.*[258] This is the equivalent of the annual consumption allowance for thirty adult male *paidaria.*[259] The *potamitai* were specialist irrigation workers who were primarily employed, however, in the low-water months, so this quantity of wine could have supported considerably more men.[260]

Another sale of wine may be attested in *P.Oxy.* XVI 2033 (VII), a text of rather difficult interpretation.[261] Line 6 reads, "and for the price of wine, 120 Alexandrian *solidi,*"[262] and since the editors report that lines 1–10 concern receipts, these *solidi* should be the proceeds of a sale. Yet lines 7 and 8 record sums "for the field guards of Saso Kato" (ὑπὲϱ τῶν ἀγϱοφυλάκων Σάσω Κάτω) and "for the expenses of the *meros* [fiscal share?]" (ὑπὲϱ ἀναλωμάτων τοῦ μέϱους), respectively, and it is difficult to understand these as income; presumably the monies in question were collected for the purposes specified. This, in turn, suggests the possibility that line 6's *solidi* were earmarked for a *purchase* of wine. The *epoikion* Saso(u) Kato, which is mentioned in three entries, is known to have belonged to the Apion estate,[263] but it is noteworthy for collecting its own taxes; for a time at least, it seems not to have been part of a *prostasia.*[264] Whatever the administrative status of the entity that produced 2033, the fact that many of the account's entries, including the one concerning the wine, are booked in Alexandrian *solidi* suggests transactions—perhaps fiscal ones—outside of the estate's *internal* monetary system (for internal transactions, we would expect the private standard).

To summarize, there is but one unequivocal sale of wine—and it is a small-scale affair—among the hundreds of documents preserved in the Apion

256. Cf. *P.Lond.* V 1773.8; *P.Oxy.* VIII 1131.5, XVI 1973.8.

257. E.g., lines 15, 39, 42.

258. On the assumption that the *solidi* in 953, an internal account, are also on the private (estate) standard.

259. Cf. p. 132 above.

260. The extent of the *oikos*' irrigation obligations is well illustrated by *P.Oxy.* XIX 2244.

261. Some interesting entries in the unedited first column may be noted for specialists:

Line 14: π]ϱ̣ο̣θμίου	νο(μ.) ζ κ(εϱ.) ιβ	(cf. *P.Oxy.* XVI 1917.110)
Line 20:] κελλαϱικ( )	νο(μ.) ς κ(εϱ.) ιη	
Line 21:] σαμαθ( )	νο(μ.) κε κ(εϱ.) ι̣β̣	(cf. appendix C, s.v. σάμαθον)

262. καὶ ὑπὲϱ τιμῆς οἴνου Ἀλεξανδϱείας νομίσματα ϱκ.

263. *P.Oxy.* XVI 1917, LXII 4350.

264. See above, p. 49 n. 60.

dossier. A handful of other references—two of them involving considerable sums of money—remain ambiguous: they may represent sales, or they may represent purchases. Of these indefinite testimonia, the best cases can be made for *P.Oxy.* XVI 2032.62 and 64, and if we accept these as sales, the fact that an employee is the buyer (i.e., is functioning as a middleman) is noteworthy: the estate has shifted away the burdens and risks (and potential rewards) of the market. Given the fragmentary state of our evidence—the huge numbers of transactions that have been lost—the dearth of data for sales should not be considered a decisive blow against arguments for market-driven production of wine. But, as we shall see in the next chapter, its stands in good accord with the conclusions suggested by other evidence.

Chapter 5

Assessing Estate Viticulture

The preceding four chapters have laid the groundwork for an assessment of the role of viticulture on the Apion estate. They have demonstrated that viticulture was the only agricultural sector through which the estate could have engaged directly with the market in a meaningful fashion. The problems surrounding the identification of estate vineyards and important viticultural terminology—frequently misunderstood, even misused (cf. Sarris on *chōrion*)—have been elucidated, and an extensive role for the estate in the management of *imperial* vineland has been posited. A descriptive approach was adopted toward estate labor; in the face of recent claims for the primacy of wage labor and direct exploitation, this process revealed the dominant presence of viticultural tenancy in three forms (leases employing cash, fixed-rate in kind, and share rents) and argued for a limited role for the *enapographoi,* whom some have regarded as the "mainstay" of the Apiones' labor force. The immediately preceding chapter illuminated the purchase, expenditure, and sale of wine. The evidence for the last of these is sparse, and this, when considered alongside the limited contribution of wage labor, suggests a similarly attenuated role for market-driven wine production. The present chapter aims to come to a firm conclusion concerning this issue and to complete another task set at the outset: the determination of viticulture's contribution to the debate about the great estate.

Ginetai Homou (or Balancing the Books)

The wine expenditures of the estate come rather close to balancing with the *hypodochē*, with the sum total of wine available to the *oikos,* in one *oino-*

cheiristēs' account (*P.Oxy.* XXVII 2480),[1] and this may also have been the case in another of these *logoi* (*P.Oxy.* LVIII 3960).[2] In the third of the well-preserved stewards' accounts, *PSI* VIII 953, there is the potential for a marketable surplus, but caution is in order because purchases of wine appear in the text.[3] It may well be that the estate pursued the same "just enough" strategy that is attested in the grain sector.[4] This would not have been difficult: a large portion of the wine expenditures detailed in the preceding chapter would have been a relatively stable cost; only under extraordinary circumstances—such as the extended presence of external military forces or the death of the *geouchos*—would the steward have needed to draw on other sources of wine. The ability to convert money rents,[5] moreover, would have afforded managers some flexibility to adjust production to meet the demand from expenses.

The poor state of the evidence for the marketing of wine has already been noted. This need not, however, deter an estimate of how much wine could have contributed to estate income. The largest volume of wine attested in the dossier is the ca. 36,700 *dipla* in *PSI* VIII 953.[6] Although this is the sum of the outflows (*analōmata*) in that text[7]—and despite the presence of purchases in 953—for the purposes of the present illustration, let it be the maximum volume of wine produced by the estate. As was made clear in the preceding chapter, not all of this wine would have been sold for profit. For the Appianus estate—arguably the paragon among market-oriented rural enterprises—Rathbone has suggested a profit-generating surplus in wine of 20–25 percent.[8] If we take the high end of this range, surplus Apion production amounts to 9,175 *dipla*. In 953.57, the tariff is 30 *dipla* to a *solidus*; if this is applied to our surplus, the net result is a profit of just under 306 *solidi*. At most, one might double this.[9] During the floruit of the *oinocheiristēs* Pamouthis, in the 540s, the

1. See p. 97 n. 20 above.

2. In the extant portion of 3960—the sections concerning expenditures and arrears are incomplete—approximately 85 percent of the *hypodochē* is consumed. The editor has commented, "It looks as if the figures would have been very close to balancing if the text had survived in full" (*P.Oxy.* LVIII, p. 119).

3. As indeed they do in *P.Oxy.* LVIII 3960. See table 4.2.

4. See p. 29 above.

5. See p. 78 above.

6. As noted above (p. 124), these were *dipla* of 8 *sextarii* (the largest size attested).

7. See lines 1–5; the grand total is 36,700.58 *dipla* (1,343 of which were vinegar).

8. Rathbone 1991, 297 (for the estate as a whole).

9. I assume a market ceiling lower than the "prices" in sales on delivery. In contemporary Oxyrhynchite contracts of this kind, prices begin at 1 *solidus* per ca. 520 *sextarii;* cf. Kruit 1992a, 183. In any case, the practice of selling to middlemen (which I discussed in chapter 4) would not have yielded the best prices.

estate expected to gross 10,743 *solidi* (on the private standard) from its rural holdings;[10] the profit from wine sales would have accounted for 5.7 percent of this *at most.*[11] The total value of all wine produced, according to the tariff in 953, would have been 11 percent of anticipated gross money income.

It has already been demonstrated that wine was the only agricultural sector in which the estate could have had significant direct engagement with the market. Even if we grant a generous marketable surplus—a surplus not supported by any extant evidence—it is clear that wine's contribution to estate income would have been marginal at best. The remainder of the money that the estate expected to receive—or, at least, a significant majority of it—must be attributed to cash rents.[12] There is no way around this.[13] That cash was king on the Apion estate is not controversial: this preference has been noted since Hardy's day,[14] and, indeed, it should come as no surprise given the centrality of the stable *solidus* in the rise of the new bureaucratic elites.[15] Yet the principal source of this cash was not the market, at least not directly, but tenancy. For direct market engagement, we must look within the estate, to those who were compelled to pay the estate's cash rents.

The issue of growth on the estate is a difficult one, but it has likely been overstated in recent literature,[16] and there is certainly no evidence for the expansion of estate wine production.[17] Some of the gold extracted from estate tenants was no doubt reinvested locally, but most of the surplus flowed to the center, to Constantinople, the focal point for family interests and aspirations, where it served as a catalyst in the production of status. Through the prism of

10. For this sum (from *P.Oxy.* XVI 1918v), see Hickey 2008, 93; it is inclusive of taxes. The comparable figure from ca. 586/87 is 18,512 *solidi,* 191.75 *keratia* (on the Alexandrian standard). On the assumption that 953, an internal account, employs *solidi* on the private standard, I have converted the Alexandrian *solidi* of 1918 (14,325.19 *sol.*) using the formula in Maresch 1994, 43. The percentages given in the paragraph above would be even lower if the interpretation of the gold standards in Banaji 1998 (which I prefer) were followed.

11. I.e., at a market price double the tariff in 953.

12. The contribution of cash rents is reduced if we allow that the estate collected taxes from property that it did not own, i.e., if we accept the "Gascou thesis."

13. This is a further indication that the role of *enapographoi,* who were typically wage laborers (cf. pp. 83–85), was limited; their predominance in the documentation is clearly a function of the fiscal character of the status. Neoclassical modeling (cf., e.g., Conning 2003) suggests that a relatively closed land market resulting from the concentration of property in the hands of a cohesive group of elites, alongside this group's access to expensive technology and economies of scale, squelches tenancy. This result is certainly not observed in the late antique Oxyrhynchite.

14. Hardy 1931, 100.

15. A central theme in the work of Banaji.

16. For discussion, see Hickey 2008, 96–100.

17. Pace Banaji 1999, 204–6.

neoclassical economics, this deployment of resources might seem "irrational," but it certainly had an objective: to keep the flow of gold from Middle Egypt coming. Investment and direct marketing had the potential to bring higher profits, but with them came expense, effort, and, worse, risk. Cash rents, in contrast, were stable, and from the landlord's standpoint, they were "good enough."[18] Risk aversion and the pursuit of stability may be considered hallmarks of the Apiones' Oxyrhynchite estate; they are the leitmotif running through its extant economic records, present in the investments made in perennial irrigation, in the control of labor, in the selling of surpluses to middlemen, and so on ad infinitum. The presence of such a "primitivist" mentality, moreover, should not cause us wonder: the literary sources that are at the foundation of the primitivist paradigm are products of the same social stratum in which the Apiones moved. Though the newer Constantinopolitan senate was in many ways removed from that of the early empire, it maintained the old social patterns.[19]

What about the *Autourgia?*

Sarris has argued that the Apion estate was "bipartite," that it consisted of two distinct sectors: the *ktēmata* and the *autourgia,* or "in hand."[20] "Ktematic" revenues were marginal; the principal purpose of the *ktēmata* was to provide revenue and manpower (through labor dues) for the market-oriented and directly exploited "in hand."[21] Because the *autourgia* figures so little in our extant sources,[22] Sarris's implication is that evidence for the estate qua economic powerhouse has been lost. On its face, this is dubious, for if the *autourgia* was well integrated with the *ktēmata*—and the accounts reveal that it was, at least economically[23]—it would seem more appropriate to assume that the texts' silence is indicative. Indeed, we might expect the *autourgia* to be overrepresented in our evidence, given the central bias of the texts that have come down to us.[24]

Fortunately, this need not remain a matter of speculation. Sarris has overlooked a central account of arrears—one encompassing both of his estate sectors—that suggests that the *autourgia* was responsible for only about 15

18. Cf., e.g., Kehoe 1997, 20.
19. Cf. Cameron 1993, 90.
20. This model has its origin in medieval economic history; cf. Sarris 2004a, 279–80.
21. Cf. Sarris 2006, 34; see also 121 n. 29 (in direct criticism of Hickey 2001).
22. For the references to it, see Mazza 2001, 129–34.
23. Cf. pp. 26–27 above, where it is also observed that the *autourgia* may have been part of the *ktēmata* (i.e., administered by *pronoētai* in *prostasiai*).
24. Cf. pp. 19–20 above.

percent of *oikos*-wide delinquencies.[25] Admittedly, this figure is open to various interpretations, but certainly one of them is that the *autourgia* was proportionally (or more than proportionally) marginal in its contribution to estate income. Other references to the "in hand," however, are more telling. By far, the best-attested produce of autourgic land is fodder, hardly the archetypical cash crop; on the contrary, this is suggestive of production for estate consumption (by animals for transport and irrigation) and fiscal purposes (*capitum*, i.e., fodder rations for soldiers and officials, as well as distributions for animals of the public post). On the basis of *P.Oxy.* LV 3804.267–69, which documents the purchase of irrigation animals for "the landlord's *autourgia* Outside the Gate," Sarris seems to believe that the entire *Exo tes pyles* was exploited directly,[26] and he adduces *P.Oxy.* XVI 1913, an account that includes expenditures for this place, as a witness to arboriculture and viticulture in the *autourgia*[27]—despite the fact that the only explicit reference to *autourgia* in 1913 (ll. 36–39) concerns fodder. But let us grant the reasonable proposition that an area of intensively cultivated orchards and gardens was present in the vicinity of the estate house. It is quite another matter to extend this model to other parts of the estate. Where is the evidence? Meanwhile, in the catalog of autourgic arrears in *P.Oxy.* XVI 2032.68–90, not a single entry explicitly concerns market crops; we are in the world of *sitos, krithē,* and *chortos,* the realm of consumption and taxes.[28]

From *Ampelos* to *Oikos*

We can next turn to viticulture's contribution to the debate about the great estate, to the controversy surrounding the so-called Gascou thesis. Chapter 4 has already given us a critical piece of information, a "red flag" for those who would write the economic history of the estate. That data point is 572, the maximum number of permanent employees suggested by the estatewide wine distributions in *P.Oxy.* LVIII 3960. Though this figure is certainly too high,[29] we can use it to calculate the number of permanent employees per *aroura*

25. *P.Oxy.* XVI 1918r.15 indicates that the estate's total arrears (in 541?) were 1,006.31 Alexandrian *solidi.* The *autourgia's* contribution to these (l. 26) was approximately 15 percent.

26. Perhaps this is due to the translation in the editio princeps, which has "the landlord's *autourgia* (called) Outside the Gate." But *Exo tes pyles* could be locative.

27. Sarris 2006, 33–34.

28. For additional criticism of Sarris's bipartite model, see Ruffini 2010, 633–35 (Ruffini seems, however, not to be aware of the correctives in Benaissa 2007, 80–86).

29. See p. 133 for further discussion.

under the traditional "large estate" model; the result (572 individuals/ca. 112,000 *ar.*) is .005 (.019/ha). This is inconceivably low[30] and stands as an impediment for Jones's area estimate (to say nothing of Sarris's assertion that Jones's figure is conservative). Assuming a smaller estate of, say, 25,000 *arourai* yields a more palatable figure (.022 or .083/ha), but it is still one that is inconsistent with an estate given over to the direct exploitation of labor-intensive crops like vines; instead, it suggests extensive use of leases and limited direct involvement in intensive forms of agriculture—precisely the exploitation "profile" suggested by production figures analyzed above.[31] Of course, it is possible to posit the existence of permanent employees whose distributions (if any) are not documented in the account of the *oinocheiristēs* (i.e., that our figure 572 is too low). Sarris's argument that the residents of the *ktēmata* were compelled to perform general labor service on the *autourgia* might be cited in this regard. Yet even if we grant a bipartite estate, there is a problem with the suggestion that individuals performing autourgic labor did not receive distributions from the wine steward. Sarris has indicated that they were *paramonarioi,* and individuals with this status are among the recipients specified in *P.Oxy.* LVIII 3960.[32] Indeed, Sarris cites 3960 for distributions to the "servile" component of his autourgic workforce, the *paidaria;*[33] that is, he himself does not believe that there was a separate (now missing) wine distribution account for those working the *autourgia.*

Estimating the area of the estate under vines confirms that it must have been smaller than Jones suggested, that Gascou's model of *oikoi* sharing fiscal and administrative functions is the correct one. Again, we can start with the largest volume of wine mentioned in the Apion dossier: the 36,700.58 *dipla* of

30. Some modern data points: An International Labour Office report (Geneva, 1969) prepared for the United Arab Republic (then just Egypt) used .057–.514 workers per hectare as its lowest category ("very low") of permanent labor per unit of area (p. 56). As a result of mechanization, one might, prima facie, expect agriculture in the modern period to have lower ratios than those from antiquity, but in Egypt's case, at least, any reductions were probably more than offset by the much smaller size of holdings (the result of land reform), among other factors (e.g., increased yields). Permanent labor input per hectare on plantation crop estates in Indonesia (1983) was .3 (.48 on Java). Smallholders had significantly higher ratios; in the coffee sector, e.g., the smallholder ratio was 6.31 (10.31 on Java). See International Labour Office, *Recent Developments in the Plantation Sector* 10 (Geneva, 1994), 44.

31. Cf. Columella (VI.3.8) and Pliny (*NH* XVII.215) on Italian viticulture: .567 and .397 workers per hectare, respectively. (These ratios refer only to the permanent labor needed to work the vineland itself; in contrast to the Apion figure, they do not include permanent support staff.) Of course, the smaller the percentage of land devoted labor-intensive crops like vines, the lower their impact on the labor input ratio for the whole estate.

32. Sarris 2006, 87–88.

33. Sarris 2006, 40.

analōmata in *PSI* VIII 953.5.[34] For the sake of argument, let us assume that the receipts and expenditures in this account came close to balancing, as they certainly did in *P.Oxy.* XXVII 2480 (and as probably was the case in *P.Oxy.* LVIII 3960);[35] a reasonable estimate for the steward's *hypodochē* seems 37,000 *dipla.* After adding a percentage—say 5 percent—for losses, we should deduct the 2,000 *dipla* known to have been purchased;[36] the resulting (production) figure is ca. 36,950 *dipla.* If we apply a yield of 33.5 hectoliters per hectare to this volume,[37] the result is an area of 48.2 hectares, or roughly 175 *arourai.* We can be confident that the *oinocheiristai* were responsible for the vast majority (if not all) of the wine received and expended by the Apiones' Oxyrhynchite *oikos*[38] and that, as a rule, there was but one of these managers in office at a time.[39] Rather more difficult are the money rents, which I have already indicated were a significant component of estate income.[40] In chapter 2, I argued that land under *apotakton chōriōn* was probably imperial, so this need not be counted.[41] But these apotactic rents might be used to establish an estimate for the area of estate vineyards yielding cash rents. The apotactic rents preserved in table 2.1 correspond to approximately 90 *arourai.*[42] These 90 *arourai* represent seven different *prostasiai,*[43] but there were certainly more than seven such districts in the *oikos. P.Oxy.* XVI 2032—the text at present furnishing the best information—suggests that there were sixteen (in 541);[44] accordingly, we can estimate about 205 *arourai* under *apotakton chōriōn,* a rent of 615 *solidi* on the public standard.[45] In chapter 2, I also suggested the working hypothesis that imperial rents were typically booked in *solidi* on the public standard.[46] For the purpose of the present illustration, let us assume that this proposition is true—

34. As we have seen, these were *dipla* of 8 *sextarii.*

35. See pp. 146–47.

36. Regarding the losses, cf. Rathbone 1991, 247 n. 51. A larger quantity was actually purchased (cf. table 4.2.4–5); I use the round number for the sake of convenience.

37. For this figure, cf. pp. 74–75 above. This yield is about 10 hectoliters per hectare greater than the (Ptolemaic) one calculated in Clarysse and Vandorpe 1997, 70.

38. Cf. p. 151 above regarding distributions to the workforce of the *autourgia.*

39. See p. 98.

40. See p. 148.

41. See pp. 57–58.

42. I do not count Nekonthis twice (cf. table 2.1.1, 17), but I do accept Ast's suggestion regarding *PSI* VIII 954.42 (for which see the notes to table 2.1). The area is determined using the rent rate in table 2.1.7.

43. Seven *prostasiai:* five distinct pronoetic accounts appear in table 2.1; to these I add one *prostasia* for Pakerke (2.1.16; for its *pronoētēs,* cf. *P.Oxy.* XVI 2034.14) and one for Pempo (2.1.7, 8).

44. So Ruffini 2008, 107 (against Hardy 1931, 82).

45. At the rate in table 2.1.7.

46. See p. 58 n. 99.

it is certainly true in part—and that no other kinds of *lēmmata* were reckoned on this standard, that is, that public *solidi* are a marker for imperial income. This would mean that *apotakton chōriōn* furnished about 20 percent of imperial land's contribution to the *lēmmata* of the *prostasiai* ca. 542.[47] If this percentage is valid for the estate's own contribution to the *lēmmata* of the *prostasiai*,[48] about 650 *arourai* of vineland under cash leases are indicated,[49] for a total (rents in kind + cash rents, i.e., 175 + 650) of 825 *arourai*. Several adjustments to this estimate are possible,[50] but 1,250 *arourai* in total (rents in kind + cash rents) seems a reasonable ceiling.[51]

A total vineyard area of about 825 *arourai* (ca. 225 ha, 555 acres) is certainly not a negligible amount, but the Apiones were one of the most powerful families in the later Roman Empire, and their estate at Oxyrhynchus has generally been considered the basis for their wealth.[52] The estimate seems rather anomalous when considered alongside Jones's great estate of 112,000 *arourai*: it would account for just .7 percent—1/135th—of such an estate, and it is inconceivable that a family of such power and wealth would have invested so little in capital-intensive agriculture, far less than those of lower means.[53] Such

47. To obtain this percentage, divide the apotactic *solidi* by the public-standard *solidi* in *P.Oxy.* XVI 1918v.6. For the *lēmmata* of the *prostasiai*, see the discussion in Hickey 2008, 89–93.

48. It has already been suggested that the Apiones imitated imperial estate management practice; cf. p. 81 above.

49. Calculation: Subtract the imperial contribution (public *solidi* in 1918v.6, converted to the Alexandrian standard) and taxes on the *lēmmata* of the *prostasiai* (1,237 Alexandrian *sol.* = total tax in 1918v.34 – village taxes in 1918v.12) from 14,325 Alexandrian *solidi* (the total *lēmmata* of the *prostasiai*; cf. 1918v.12); multiply the remainder by .2, then divide by 3.083 *solidi* per *aroura* (table 2.1.7's vineyard rent converted to the Alexandrian standard).

50. A larger apotactic area (necessitated because, e.g., the data in table 2.1 are not representative, or there were more *prostasiai*) would increase the figure, as would, of course, a greater percentage contribution by cash vineyard leases to the estate's *lēmmata*. If imperial rents were not the sole contributor to the *lēmmata* in *solidi* on the public standard, the percentage that I have used as a guide would also be higher. (Shortly before this manuscript was submitted to the press, Amin Benaissa sent me his transcription of an unpublished account of grain arrears that indicates that there were, in fact, more *prostasiai*, at least ca. 575. By my calculation, however, the area that I have proposed herein as a ceiling still holds.)

51. 1,250 *arourai* ≈ 19.25 *arourai* at every *epoikion*-level (i.e., "estate") site in appendix B. Areas of ca. 30 *arourai* are attested for *epoikia* as a whole (cf. Bagnall 1993a, 151 and n. 17). By way of comparison, there were 10.5 *arourai* of vineland at the *epoikion* Nekontheos (Text I), 3 at the *epoikion* Phankona (Text I), and 4 at the imperial *ktēma* Pempo (*P.Oxy.* XVI 1915).

52. Cf. *ODB*, s.v. "Apion."

53. Cf., e.g., Kehoe 1992, 118: "The greater resources that an Appianus had at his disposal in comparison to more modest landowners allowed him to engage to a greater degree in producing crops, such as wine, that required a greater investment of labor and capital, and it is likely that wine provided a greater portion of his income from his estates than it did for more modest landowners."

a conclusion stands even if we employ the ceiling that I have suggested, that is, 1,250 *arourai* (which equals a mere 1.1 percent).

Rostovtzeff cites *P.Amh.* II 79, a petition to the *praefectus Aegypti* from the late second century, as an example of the immense fortunes that certain members of the "local bourgeoisie" in Roman Egypt were able to acquire.[54] One of the "villains" in this text is said to have illicitly accumulated 7,000 *arourai* of farmland and 200 *arourai* of vineland; vineland therefore amounted to 2.8 percent of his agricultural land. If this figure is applied to our vineyard estimate, we arrive at an Apion estate of 29,500 *arourai* (ca. 81 km^2), nearly four times smaller than the area calculated by Jones. Of course, the patrician Apiones hardly qualify for membership in Rostovtzeff's "bourgeoisie," and, as implied above, the percentage of vineland on their estates may well have been higher. The higher the percentage, the smaller the estate: 3.75 percent would correspond to 22,000 *arourai* (ca. 61 km^2), 5 percent to 16,500 (ca. 45 km^2), 7.5 percent to 11,000 (ca. 30 km^2), and so forth.[55] The first figure in this sequence, it may be noted, puts us "in the ballpark" with my estimate—derived from independent data—of estate arable land: 18–22,000 *arourai.*[56] In any case, the central tenet of Gascou's model of fiscal participation has been vindicated; there is no other way to reconcile these reduced area figures and the 112,000 *arourai* implied by the fiscal documents that Jones employed. The latter area can only correspond to a fiscal share, the sum of Apion lands and the holdings for which the family collected taxes. It is only when we include these non-Apion holdings in the 112,000 *arourai,* when we dispense with the large feudal estate, that we can begin to reckon with 825 (or 1,250) *arourai* of Apion vineland.

There is a further ramification to this "small estate" model. It is an ancient historians' chestnut that eastern senators were much less wealthy than their western counterparts,[57] but the Apiones are generally regarded as the exception par excellence.[58] Though findings concerning a single node of the family's holdings must be considered preliminary—and, in any case (as Apion III's marriage to Eusebia demonstrates), the boundaries between the two

54. Rostovtzeff 1957, 296.

55. Although Rostovtzeff (1957, 296) regards *P.Amh.* II 79 as "probably trustworthy as regards the extent . . . of the fortune," the context suggests the possibility of exaggeration. One might expect the more expensive vineland to be most inflated; if this was the case, the actual percentage of vineland would have been smaller than 2.8 percent. Application of a smaller percentage to my vineyard estimate would result in a somewhat larger Apion estate, but the difference in wealth between the Apiones and the individual in *P.Amh.* II 79 makes such a small percentage rather unlikely.

56. Hickey 2008, 91.

57. Cf. Jones 1964, 554–56.

58. Cf., e.g., *CLRE,* 12; Finley 1999, 102.

groups were certainly permeable—the figures proposed here at least open the door to the possibility that the Apiones were more typical members of the eastern service aristocracy that filled the "created" senate in Constantinople.[59] Even the presumed expenses of Apion II's consulship cannot be cited against this: there could have been imperial assistance, a proposition that is quite plausible—as is receipt of the ordinary consulate in the first place from a "skittish" Justinian—in the context of the family's close relationship with the emperor.[60] In any case, to compete with the super rich families based in the heartland of the old Roman Empire,[61] the Apiones would have needed significantly more wealth than their estate at Oxyrhynchus could possibly have generated. The source of such income remains to be discovered.

59. For the formation of the eastern senate, cf. Zuckerman 1998, 121–35.

60. Cf. *CLRE*, 10–12. The silence about this relationship in some scholarly literature continues to baffle me.

61. Cf. Wickham 2005, 162–63, citing the well-known evidence (e.g., the 120,000 *sol.* that Melania or Pinianus received annually as rents).

Conclusion

By virtue of its extensive documentation, the Apion estate has long been at the center of arguments concerning the economy and society of late antique Egypt; indeed, it is one of the few economic entities that has managed to transcend local—in this case, we might say "papyrological"—history, to enter the historiography of the wider world of the Byzantine East and beyond.[1] This is certainly a good thing, for papyri *as a corpus* offer unparalleled opportunities for "mainstream" history, possibilities that have hardly been exploited fully. Rather more remarkable than this wide currency, however, is the estate's apparent mutability. One scholar can interrogate the sources and find a "glorious house" that is medieval; another, something eminently modern and "rational." To one it is a force pulling the state apart, even partly responsible for the success of Islam; to another, it is helping to prop up the empire. Wage laborers and slaves dominate one estate, tenants another. The published evidence, in the meantime, has remained largely unchanged for decades; we have gained more details, of course, particularly in the last decade, but the trends of the sources remain much the same. There is an interesting hermeneutic phenomenon here.

The aim of this book has been to hem in Proteus, to limit possibilities. It has not sought to provide firm answers, for even the papyri, for all their richness, are rarely capable of offering such. But they do permit us to establish boundaries. Though models-based history lies at the root of our interpreta-

1. Of course, the rank and fame that the Apiones obtained have contributed to this transcendence, as has the presence of most of the texts in the more accessible *Oxyrhynchus Papyri* series.

tive difficulties, my purpose is not its indictment; the study of antiquity is much diminished without the presence of interpretative frameworks. I would contend, however, that some sort of dialectic between evidence and model is essential. The present work represents "push back" by the evidence.

A key construct under consideration here has been that of the "great estate." The first attempt to quantify the "greatness" of the Apion estate, the prototypical great estate, came from A. H. M. Jones, who indicated that it comprised 112,000 *arourai*, a substantial percentage of the total area of the Oxyrhynchite and Kynopolite nomes. This was countered in the mid-1980s by Gascou, who proposed that the glorious house was a willing administrative partner of the state, that 112,000 *arourai* simply indicated the *oikos*' share of fiscal responsibility; the estate itself was a smaller entity. Recently, Sarris has attacked Gascou, arguing that the Apion estate was, in fact, considerably larger than Jones proposed because Jones did not include the area of the *autourgia*, the estate's directly worked land. At present, Sarris's position has wide acceptance among historians of late antiquity. The matter is of broader significance because of its historiographical implications, which are amplified by the centrality of the Apion estate to the discourse: if we eliminate the large Apion estate, feudal (*sensu lato*) conceptions of late antique Egypt are significantly weakened.

A 2008 publication of mine already called the "great estate" into question, and I have sought independent corroboration of this in the present work. The approach here is different. In 2008, a single text sufficed, but such low-lying fruit are rare. In this volume, I have had to reconstruct a sector of estate production in order to lay the groundwork for a rough calculation of the area devoted to it; using that calculation and comparative data, I, in turn, have delimited the area of the estate itself. *Pour faire coup double,* I chose to analyze viticulture, since recent commentators have indicated that it was the focal point for estate investment and was characterized by market-driven production employing wage labor. As was demonstrated in chapter 1, wine is, in fact, the only production sector in which the estate could have had meaningful direct engagement with the market.

We can attribute the necessity for a painstaking reconstruction of estate viticulture to the difficulties (fragmentation, bias, etc.) of our sources, as well as to their apparent susceptibility to equivocal readings. Chapter 2 accordingly began with fundamentals, addressing key and often misinterpreted terms from the vineyard of the late antique Oxyrhynchite. I argued that the definition "vineyard" should be the *point de départ* for *chōrion,* while I defined *meris* as a "section of a vineyard." This chapter also illuminated the poorly understood aide-mémoire *dikaion* and documented the ambiguity of ownership, culminating with the suggestion that *apotakton chōriōn* in the Apion accounts indicates imperial vineland under estate administration. The ultimate goal of chapter 2, besides clarification (and, in some cases, problematization), was the creation of a list of every settlement connected with estate viticulture (see appendix B).

Chapter 3 turned to labor questions. Banaji and Sarris have emphasized the role of wage labor on the estate, but in viticulture—the sector in which we would expect direct exploitation to be most pronounced—it is quite clear that tenancy was dominant. Fixed-rate rents in both cash and kind are attested, as is sharecropping (hitherto thought moribund in the Byzantine Oxyrhynchite). Wage labor is primarily documented in irrigation, and yield data suggest that tenants paying rents in kind were usually joined with an irrigator who worked for the estate. Of course, the distinction between certain forms of tenancy and wage labor can be small. As for the *coloni adscripticii* or *enapographoi* (who were typically, but not exclusively, wage laborers), they were hardly, as Sarris has insisted, the mainstay of the Apiones' labor force. Pronoetic accounts suggest that they may have been confined to certain settlements, and even if they were present throughout the estate, their numbers were limited. The fiscal character of the status has clearly enhanced their visibility.

The flow of wine through the estate was the focus of chapter 4. By virtue of extensive and well-preserved accounts of the estate wine stewards, or *oinocheiristai,* we are very well informed about the disposition of estate wine stocks. Expenditures for pious and charitable causes, to members of the civil administration and military units, for internal expenses like salaries, and in connection with festivals and spectacles are documented in remarkable detail. The testimonia for sales, in contrast, are truly limited and, in most cases, equivocal, and there is clear evidence that the estate purchased wine. That the wine steward was the manager most likely to engage in the wine trade on behalf of the estate (and that there was, as a rule, only one steward at a time) cannot be doubted.

Chapter 5 assessed the importance of viticulture to the estate. Although sales are very poorly attested, I nonetheless assumed a generous marketable surplus in wine of 25 percent. Under this unlikely model, it is quite clear that wine's contribution to anticipated gross income still would have been rather limited: *at most,* wine sales would have contributed less than 6 percent. Cash rents from tenancy were expected to supply the bulk of the remainder. There is no reason to think, as Sarris has argued, that the production from the *autourgia* has gone undetected; the principal crop of this directly worked land was clearly fodder, no doubt intended for internal consumption and taxes.

Chapter 5 concluded with an analysis of indicators that are relevant for the area of the estate: the size of its permanent labor force and a calculation of its area under vines. The former suggests that Jones's estimate is inconceivably large and supports the argument that tenancy predominated on the estate. The vineyard area estimate similarly confirms a small estate, and I conclude that the central tenet of the Gascou model—that *oikoi* participated in the fiscal administration of the Oxyrhynchite on a system of shares—must stand.

Some questions do, however, remain. By its very nature, the Apion dossier is a limited source for the mentality of the estate's owners and managers, for the *process* of decision making. I suggested above that "old" Roman aristocratic values still prevailed and that the Apiones were satisfied with a "sufficient" and stable supply of Egyptian gold to deploy in the Constantinopolitan status economy, but this issue would bear further exploration (one would want to include other aristocrats and draw on papyrological and nonpapyrological sources). Similarly, while "fiscal participation" by the estate is unquestionable, the motivation for this (to say nothing of its genesis in the fifth century) deserves additional consideration.[2] "Was it willing or unwilling?" has been one way to formulate the question, but this lacks nuance. Doubtless, the posture of the owners was to some degree dependent on that of the emperor and on the closeness of their relationship (cf. the situation under Justinian and Anastasius). The Apiones, moreover, were not the only participants in the system, and we should expect a diversity of responses and motivations; not everyone had the emperor's ear. Finally, though "willing participation" need not have the good of the state as a motive, I do think that it is a mistake to discount the role of euergetism. This whole discussion, which is critical to the narrative of state formation, demands more subtlety.

Applicability is yet another issue needing attention. How typical of the Apiones' holdings was the Oxyrhynchite estate? Were other units managed more aggressively? This we will probably never know, but the estate's typicality within the Oxyrhynchite and Egypt as a whole is another matter. My forthcoming publication of the dossier of Flavia Anastasia will illuminate the economic posture of a middling Oxyrhynchite contemporary of the Apiones, and examples of smallholder behavior are well known through the Aphroditan corpus and might be documented in other texts.[3] A comparative study would be unproblematic. But was the Oxyrhynchite social landscape (and its attendant fiscal system) typical of the rest of Egypt? In his important study

2. For more on the development of the system of shares, cf. Azzarello 2006.
3. See Hickey 2007, 304–6 with references, as well as *P.Worp*, pp. 348–50.

Du village à l'Empire, Zuckerman has suggested that the large estate of Iulianos lurks behind the Aphrodite papyri;[4] his evidence, however, is scant. Ruffini, in turn, has proposed that the traditional dichotomy between "great estate" Oxyrhynchus and "free village" Aphrodite might be a deception due to the scales of their respective corpora.[5] At present, the jury remains out, though I find myself skeptical.

---

4. Zuckerman 2004, 221–22.
5. Ruffini 2008, 246–51.

Appendixes

Appendix A

*Abstract of Text I (P.Oxy. XVI 1916v)**

Key to Column Headings

A = cellar charge
B = current charges: total
C = 14th indiction arrears
D = transfers
E = *naulon ploiou* (freight charge: boats)
F = *naulon paktonariou* (freight charge: tenders)
G = *kamēlarioi* ([to] camel drivers)
H = amount delivered (without transport charges)
I = amount delivered (with transport charges)
J = current shortfall (negative numbers indicate a surplus)
K = total owed
L = number of *arourai*

TABLE A.1. 1st Site (name unknown; share tenancy)

Ampelourgos	*Geouchikon Meros*	A	B	C	D	E	F	G	H	I	J	K	L
?	298	4	302	12	0	0	2	0	299	301	1	13	not given
Biktor	379	4	383	25	0	0	3.5	0	394	397.5	–14.5	10.5	not given
Ammonios	395	4	399	15	0	0	4	0	401	405	–6	9	not given

(continues)

*This papyrus was read in full at the Egyptian Museum in Cairo during spring 1997. My inability to obtain a suitable image has prevented publication in the intervening years.

TABLE A.1—*(Continued)*

Ampelourgos	*Geouchikon Meros*	A	B	C	D	E	F	G	H	I	J	K	L
Ischyrion	321	4	325	37	0	0	3.5	0	348	351.5	–26.5	10.5	not given
Georgios	316	4	320	26	0	0	3	0	325	328	–8	18	not given
Kyriakos	294	4	298	0	0	0	2.75	0	275	277.75	20.25	20.25	not given
Abraamios	483	4	487	16	0	0	4.75	0	487	491.75	–4.75	11.25	not given
Totals	2,486	28	2,514	131	0	0	23.5	0	2,529	2,552.5	–38.5	92.5	not given

Typical entry for 1st site:
(καὶ) (πρὸς) Γεώργιον ἀμ(πελουργὸν) Μηνᾶ γεωρ(γοῦ), γεουχ(ικοῦ) μέρ(ους) δι(πλᾶ) τις ἐχθ(έσεως) δι(πλᾶ) κς καταγαί(ου/–ίων) δι(πλᾶ) δ (γίν.) δι(πλᾶ) τμς
(ὧν) Ἀθανασίῳ οἰνοχειρ(ιστῇ) δι(πλᾶ) τκε ναύλ(ου) πακτοναρ(αρίου) δι(πλᾶ) γ (γίν.) δι(πλᾶ) τκη λοι(πὸν) δι(πλᾶ) ιη

TABLE A.2. *Epoikion* Phankona (fixed rents in kind)

1st Party	Rent	A	B	C	D	E	F	G	H	I	J	K	L
?	344	0	344	19	0	6.667	0	.667	338	345.334	−1.334	17.666	not given
Serenos	172	0	172	10	0	3.333	0	.333	169	172.666	−0.666	9.334	not given
Totals	*516*	0	*516*	29	0	10	0	1	507	518	–2	27	3

TABLE A.3. *Epoikion* Nekontheos (fixed rents in kind)

1st Party	Rent	A	B	C	D	E	F	G	H	I	J	K	L
Petros	750	0	750	53.5	0	8	0	1	430	439	311	364.5	5
Maximos	300	0	300	0	0	5	0	0.5	257	262.5	37.5	37.5	2
Petros	300	0	300	0	19	4	0	0.5	230	234.5	65.5	65.5	2
Ptollas	225	0	225	0	−19	2	0	0	97	99	126	126	1.5
Totals	1,575	0	1,575	53.5	0	19	0	2	1,014	1,035	540	593.5	10.5

Typical entry for *epoikion* Phankona and *epoikion* Nekontheos:
(καὶ) (πρὸς) Πέτρον καὶ Φιλήμμωνα ἀμ(περλουργοὺς) Ἰερεμίου καὶ Φὶβ καὶ Γερμανοῦ Θέκλας καὶ Πέτρου Ἰωάννου γεωρ(γῶν) ὑπὲρ (ἀρουρῶν) β δι(πλᾶ) τ καὶ (ὑπὲρ) ἐχθ(έσεως) ιδ ἰνδ(ικτίονος) δι(πλ–) *vacat*
(ὧν) ἐδόθ(ησαν) Ἀθανασίῳ οἰνοχειρ(ιστῇ) δι(πλᾶ) σια ναύλ(ου) πλοί(ου) δι(πλᾶ) δ καμηλ(αρίοις) δι(πλᾶ) L (γίν.) δι(πλᾶ) σιεL κτλ. (There follows a transfer of wine to another group.)

Note: Only the name of the first party listed in each account entry is given in the tables. All figures (except under L) are in *dipla;* the number of *sextarii/diploun* is not specified (but cf. p. 190 below). Figures that are in outline have been restored; those that are in italics are emendations.

Appendix B

Viticultural Sites

The following is (of necessity) a working list of sites that have some link to both the Apiones and viticulture or wine production. Toponyms, if declinable, are presented in the genitive because this is the form that almost always appears (and generating the usually unattested nominatives can occasionally be problematic).[1] Multiword place names of the form adjective + substantive are alphabetized under their nominal (or nominalized) component. Sites are classified (e.g., *kōmē*) as they are attested during the floruit of the Apiones; it should be remembered that this designation may have been different in earlier periods.[2] Their classification is followed by their location—first their toparchy and then, if possible, their *pagus*. Though the toparchies had disappeared long before the Byzantine period,[3] their larger areas make them useful; that is, one's margin for error is greater. If the location of a site is given without comment, it means that I have followed Benaissa.[4] The sources for localizing information with other origins are given in footnotes; if a location is my own, I have provided my rationale. It should be remembered that most of these sites are smaller "estate" entities (*epoikia, ktēmata, chōria*) and thus are very difficult to place. Certain reasonable assumptions have been made—for

1. Cf., e.g., Falivene 1997, xv.

2. Cf., e.g., the case of the *epoikion*, later *kōmē*, Dositheou; see Rowlandson 1996, 15. Krüger (1990, 43–44) is not alone in suggesting that the distinction between, e.g., *epoikia* and *kōmai* became blurred during late antiquity, but I am inclined (and not only with respect to this matter) to assume a terminological significance unless there are clear reasons to think otherwise. For Krüger's (and others') misunderstanding of the meaning of *chōrion* in the Byzantine Oxyrhynchite, see pp. 41–42.

3. The *pagi* ceased being used in the first half of the fifth century.

4. Benaissa 2009, which has just superseded Pruneti 1981 and Krüger 1990.

example, that the sites within a certain *prostasia* were close enough to one another to be administered conveniently,[5] that villagers worked land that was relatively close to their homes,[6] and so on—but when these have been employed, I have indicated such in the notes. Also included are some speculative and tangential places, but within limits. I have not listed Apion sites, for example, for which viticulture is only attested in the Roman period (i.e., before 400). This catalog's aim is not to identify every vine in the Oxyrhynchite (even if that were possible);[7] instead, I have sought to give an account of the centers of viticultural activity that are linked—even if only by a significant potentiality—to the Apiones.

An asterisk (*) before an entry indicates that it does not appear in Mazza 2001, appendix 6, or that Mazza has not noted viticultural activity at the site.[8]

ΑΔΑΙΟΥ(*kōmē;* Eastern Toparchy, fourth *pagus*)

—In *P.Oxy.* XVI 2044.3 (probably 579), the *oinocheiristēs* Phib segregates 1,500 *dipla* of vinegar that had come from the *kōmē* Adaiou; for discussion, see p. 96 n. 20 above.

—In *P.Oxy.* LVIII 3960.33 (621), the inhabitants of this place owe 225 *knidia* of wine.

ΑΠΕΛ(Λ)Η (*epoikion, ktēma;* Upper Toparchy)[9]

—*P.Oxy.* LV 3804.34 (566) reads, παρὰ τοῦ κοινοῦ τῶν γεωργῶν καὶ ἀμπελουργῶν ὑπὲρ ἀποτάκτου χωρίων, 43.5 *solidi.*[10] This entry is in the section of the account that concerns the *epoikion* Apele.

* **ΑΠΟΛΛΩΝΟΣ** (*kōmē;* probably Kynopolite)[11]

—Not a locale with Apion vineyards, but a source of wine purchased by the *oikos* (*P.Oxy.* LVIII 3960.14 [621], 864 *knid.* at 24 *knid./sol.*).

5. Cf. *P.Oxy.* LV 3804, p. 96.

6. Generally true, but cf. Bagnall 1993a, 139–40.

7. Sadly, the evidence does not allow us to arrive at conclusions like that (concerning Macedonia in the fourteenth century) in Laiou 1977, 175: "The relatively equal distribution of vineyards among a population which in other respects was economically unequal reflects a basic fact of peasant life."

8. The forthcoming dossier of the *oinocheiristēs* Philoxenos attests the following viticultural sites: Anthesma/Antisma, To Herakleion, Hexarourias, Loukiou, Neophytou Hierakos, Nikes, Ouesobtheos, Metrodorou, Megales Paroriou, Skytalitidos, Stephanionos, Tantape, and Tillonos. Amphora production is attested at Nikes, Leonidou, Ouesobtheos, Metrodorou, Neou, Megales Paroriou, and Pera Mermerthon. (I thank Nick Gonis for this information.)

9. Cf. *P.Oxy.* LV 3804, intr., pp. 96–97.

10. For *apotakton chōriōn,* see pp. 53–58.

11. For the location, cf. *SB* XVIII 13741.

* **ΑΨΕΜΨΕΩΣ** (*kōmē;* Upper Kynopolite)

—Not a locale with Apion vineyards, but a source of wine purchased by the Apiones' *dioikētēs* at Kynopolis (whether for himself or the *oikos* is unclear; *P.Oxy.* LVIII 3954 [611], two hundred Samian jars).

* **ΑΣΠΙΔΑ** (*epoikion*)

—*P.Oxy.* LXIX 4755 (586) is a receipt for a large cogwheel for a *geouchikē mēchanē* that may furnish water εἰς ἄμπελον καὶ εἰς ἀρόσιμον γῆν.[12] The recipient's *origo* is the *epoikion* Aspida. In 4755, this *epoikion* belongs to Fl. Ioannes, but in earlier years, it was a possession of the Apiones (cf. *P.Oxy.* XVIII 2204, ca. 565–66).

ΓΝΑΦΕΩΝ

See ΚΝΑΦΕΩΝ.

* **ΓΡΑΕΙΔΟΣ** (*epoikion*; Middle or Thmoisepho Toparchy)[13]

—*P.Oxy.* XVI 1912.152–55 (probably 554/55; cf. *BL* X 144) records the total price (11 *sol.* less 49.5 *keratia;* cf. table 3.2.2) of new reeds bought for newly planted *chōria* (1 *aroura* at this site).

* **ΔΑΦΝΟΥ**

—Krüger writes, "Wahrscheinlich ein Weingut der Apionen,"[14] and he cites *P.Oxy.* XVI 2042 (V). The link with the Apiones escapes me.

* **ΕΙΡΗΝΗΣ** (*epoikion, ktēma*(?), *chōrion;* Hermopolite, near the border with the Oxyrhynchite)[15]

—This is an Apion possession in *P.Wash.Univ.* II 102 (V/VI), but it appears to have left their control before 23.ix.611 (cf. *P.Oxy.* LVIII 3955, especially ll. 6–7). There is an undated potsherd, probably from a wine jar, that bears the words *chōrion* Eir( ); see *SB* I 1989f, and cf. p. 43 above. (This site is not attested before the fourth century.)

ΕΞΩ ΤΗΣ ΠΥΛΗΣ (near the city of Oxyrhynchus; refers to the *proastion Exo tes pyles,* on which see also ΠΡΟΑΣΤΙΟΥ.)

—*P.Oxy.* XVI 1913.1–5 (555?) mentions *chōria* here (appropriately

12. Regarding the ownership of land irrigated by such machinery, see pp. 50–51.

13. A *kōmē* Petne was part of the same *prostasia* (*P.Oxy.* XVI 1912.98–110), but there were two Oxyrhynchite villages with this name. The one in the Middle Toparchy (as opposed to the Upper) is probably meant in 1912; cf. the toponyms in lines 75 (Senepta, Middle Toparchy) and 103 (read Τ[ά]νεως, Middle Toparchy), as well as line 76 (Nomographou, Lower Toparchy).

14. Krüger 1990, 269.

15. Cf. *P.Oxy.* LVIII 3955, 6–7n. The site should not be confused with its Arsinoite homonym, for which cf. *SB* XX 14402–3 (where the editor is incorrect to render *chōrion* as "Dorf").

translated by the editors as "vineyards.") Line 49 mentions orchards with grapes (probably not used for wine).
—*P.Oxy.* LXVIII 4704 (626) is a *misthos* payment to *potamitai* working εἰς τὰ νεόφυτα χώρια. Though the text dates to the Persian period, the Apion estate is still functioning as an economic unit.

* **ΕΠΙΣΗΜΟΥ** (*kōmē;* Upper Toparchy, third *pagus*)
—*P.Oxy.* LVIII 3960.12 (621) records 498 *knidia* ἀπαιτηθέντα παρὰ τῶν ἀπὸ Ἐπισήμου χάριν τῶν Περσῶν.

ΕΡΩΤΟΣ (*epoikion, ktēma;* Lower Toparchy)[16]
—*P.Oxy.* XVIII 2195.71 (576/77) reads, παρὰ τοῦ κοινοῦ τῶν γεωργῶν ἀμπελουργῶν (so Ast 2002, 185 n. 25) ὑπὲρ ἀποτάκτου χωρίων, 9 *solidi.* This entry is in the section of the account that concerns the *ktēma* Erotos.

ΕΥΤΥΧΙΑΔΟΣ (*epoikion;* Lower Toparchy?)[17]
See ΤΕΡΥΘΕΩΣ.

* **ΗΡΑΚΛΕΙΑΣ** (*ktēma;* Middle or Thmoisepho Toparchy)[18]
—*P.Oxy.* XVI 1910.13 is a concession (in wheat) for a certain *topos* here that has been planted with vines.[19]

* **ΘΕΑΓΕΝΟΥΣ** (*epoikion, ktēma;* eighth *pagus*)
See ΤΕΡΥΘΕΩΣ.

* **ΙΒΙΩΝΟΣ** (*kōmē*)[20]
—Not a locale with Apion vineyards, but a source of wine purchased by the *oikos* and sent to Apion II in Constantinople (*PSI* VIII 953.72–74

16. It is clear from *P.Oxy.* XVIII 2195 (cf., e.g., l. 111) that the *kōmē* Sinary was nearby. This village is known to have been in the Lower Toparchy; the *epoikion* in question was probably in either the eighth or ninth *pagus.*

17. An *enapographos geōrgos* from the site in question appears alongside two from the *epoikion* Megalou Moucheos in a surety document (*SB* XVI 12484 [584]). This need not indicate that their *origines* were proximate (for discussion, see Fikhman 1980, 69), but if it does, the *epoikion* Eutychiados would have been located as indicated. The location of the *epoikion* Theagenous (see s.v. ΘΕΑΓΕΝΟΥΣ), which was administered by the same individual who oversaw the *epoikion* Eutychiados (cf. *P.Oxy.* XVI 2019), furnishes additional support for a northern location.

18. *P.Oxy.* XVI 1910 (late VI or VII) indicates that it was administered with the *ktēmata* Leonidou *kai* Limeniados, which were in the former Middle Toparchy, but the presence of the Herakleopolite village Koba (twelfth *pagus*) in this text suggests the possibility of something farther north (in the Thmoisepho Toparchy).

19. For the reading, see Hickey 1999, 325–26 (= *BL* XII 142).

20. The name ("Ibis Sanctuary") is ubiquitous. If this is the same place as that attested in *P.Oxy.* XVI 1917, it may have been in the Herakleopolite; cf. Falivene 1997, 202 n. 2. For other possibilities, see Benaissa 2009, 87–91.

[567/68]: 2,000 *d* + a 3 percent addition labeled ὑπὲρ ϱϱ( ); the editor suggests *roga*, but I prefer *rhysis*, i.e., *hyper rhyseōn*).

* **ΙΕΡΕΩΝ** (*epoikion;* Western Toparchy)[21]

—*P.Genova* II 72 (VI) probably refers to *ampelourgoi* of the *epoikion* Hiereon. Note also *P.Oxy.* XVIII 2197.71ff. (ca. 585?; see *P.Oxy.* LXX 4792, 10n.), where this *epoikion* appears yoked with the known Apion *epoikion* Partheniados. At 2197.75, bricks are expended for repairs on the house of an *ampelourgos*.

* **ΙΣΙΕΙΟΝ ΠΑΓΓΑ** (*chōrion, kōmē;* Upper Toparchy, first *pagus*)

—A *chōrion* with this name is attested in *PSI* III 165 (441/42), and the village was associated with the estate *autourgia* (*P.Oxy.* XVI 2032.80); note also *P.Oxy.* LV 3804.148.

* **ΚΑΛΩΡΙΑΣ/ΚΑΛΟΥΡΙΑΣ** (*epoikion, ktēma, chōrion;* Western or Eastern Toparchy?)[22]

—This *epoikion* is associated with a *mēchanē* suggestively named Τῶν Χωρίων (*P. Oxy.* XIX 2244.28 [528/543/558; cf. Gonis 2004a, 201]). The text in which the *mēchanē* is attested certainly stems from the Apion administration, but the irrigation machine itself may be imperial property.[23]

—For an undated potsherd (probably from a wine jar) bearing the name of a *chōrion* Kalour( ), see *SB* I 1989e, and cf. p. 43 above.

* **ΚΙΝΕΑΣ** (*epoikion*)

—*P.Oxy.* LXXII 4901 (408) is an advanced sale of wine (88 *d*) in which a deacon from the *epoikion* Kineas is the seller-debtor. In *P.Oxy.* 4901, the settlement belongs to the *domus divina*, but it later was an Apion property (cf. *SB* XXII 15362). For the history of its ownership, cf. 4901, 4n., and p. 52 above.

ΚΛΕΩΝΟΣ (*epoikion;* Middle or Thmoisepho Toparchy)[24]

—In *P.Oxy.* XVI 1912.137–39 (ca. 555; cf. *BL* X 144), land probably at this site receives a concession for *apotakton chōriōn* (1 *ar.*),[25] but the poor condition of the text renders the details unclear.

21. See ΠΑΡΘΕΝΙΑΔΟΣ below; also note the presence of two individuals *apo* Leukiou under the geographic heading for the *epoikion* Hiereon in *P.Oxy.* XVIII 2197 (ll. 76, 82).

22. *P.Oxy.* XVIII 2197.145 suggests that it was near ΟΡΘΩΝΙΟΥ (see s.v. below).

23. If Bonneau's resolution and interpretation of the abbreviation χειϱ( ) is correct; cf. *BL* X 148. I am not completely convinced.

24. See n. 13.

25. Cf. the remains of the text with line 153.

—*P.Oxy.* XVI 1912.152–55 records the total price (11 *sol.* less 49.5 *ker.;* cf. table 3.2.2) of new reeds bought for newly planted *chōria* (less than 1 *ar.* at this site).

ΚΝΑΦΕΩΝ/ΓΝΑΦΕΩΝ (*epoikion;* Western Toparchy)[26]

—*P.Oxy.* XVI 1913.29–33 and 51–53 (555?) indicate that wine jars were furnished to the *ampelourgoi* of this *epoikion.* The settlement predates the Apiones; cf. *P.Mich.* XI 611 (412).

ΚΟΤΥΛΕΕΙΟΥ(*epoikion, ktēma;* Upper Toparchy)[27]

Note: P.Oxy. LV 3804.127 (566) is an account heading for the *epoikion* Kotyleeiou δίκαιον τῶν υἱῶν τοῦ ἐπισκόπου (see also *P.Oxy.* XVI 1911.55 + *BL* IX 189 [557]); these sons were named Apollos and Phoibammon. Since their "*dikaia*" are mentioned only in lines 128 and 129, respectively (and entries for the settlement continue until l. 138), perhaps only part of this *epoikion* was the former property of the bishop's sons.[28]

—*P.Oxy.* XVI 1911.91 is a concession τοῖς γεωργοῖς τοῦ κτήματος Κοτυλεείου ὑπὲρ ἐνχόρτου ἀμπελῶνος, 12.25 out of 13 *arourai* at .667 *solidus* per *aroura;*[29] so also LV 3804.163.

—*P.Oxy.* XVI 1911.104–6 is a concession τοῖς γεωργοῖς τοῦ κτήματος Κοτυλεείου ὑπὲρ ἀρουρῶν γ φυτευθεισῶν ἐν ἀμπέλῳ κατὰ λιβὸς τοῦ κτήματος Κοτυλεείου, .688 *aroura,* total 4.5 *cancellus artabai,* 4 *choinikes,* and .5 *solidus;*[30] cf. *P.Oxy.* LV 3804.176–77, where the *arourai* are 5 and the concession (16.5 *canc. art.*, 7 *choin.*, and 1.667 *sol.*) is for the full area.[31]

—*P.Oxy.* XVI 1911.175–77 records the total price (2 *sol.*, private standard, with *rhopē*) of new reeds bought for newly planted *chōria* (1.25 *ar.* at this site).

—*P.Oxy.* XVI 1911.181ff. records that wine jars were furnished to the *ampelourgoi* of this locale;[32] cf. *P.Oxy.* LV 3804.218–19 (where they receive a portion of eleven hundred used wine jars).

—*P.Oxy.* XVI 1911.204 is a concession for 1.5 *solidi* to the *ampelourgoi* of Kotyleeiou for lands scoured away and destroyed by the breach of a dike.

26. It appears in the same *pronoētēs'* account as *Exo tes pyles;* for discussion, see p. 27 n. 36, p. 167 above.

27. Cf. *P.Oxy.* LV 3804, intr., pp. 96–97.

28. For *dikaion,* see pp. 58–60 above.

29. For the reading, see *BL* IX 190.

30. For the reading, see *BL* XI 154.

31. For discussion, see LV 3804, 176–77n.

32. For corrections to this section of the account, see *BL* XI 154–55.

—*P.Oxy.* LV 3804.135 reads, παρὰ τοῦ κοινοῦ τῶν γεωργῶν καὶ ἀμπελουργῶν ὑπὲρ ἐδάφους τῶν ἀμπελουργῶν, 13.5 *solidi.*
—*P.Oxy.* LV 3804.164–67 includes a grant of 2 *solidi* from an *edaphos* Ἔρωτος καὶ Ἀμάτου καὶ τῶν ἀμπελουργῶν in the *ktēma* Kotyleeiou; cf. also *P.Oxy.* XVI 1911.92–95 + *BL* IX 190, where καὶ τῶν ἀμπελουργῶν is apparently replaced by καὶ τοῦ ἄλ(λου) ἐργ(άτου) τῶν ἀμπελ(ων).
—*P.Oxy.* LV 3804.191–92 is a concession Νῶε καὶ Πέτρῳ ἀπὸ Κοτυλεείου ἀμπελουργοῖς τοῦ αὐτοῦ κτήματος Κοτυλεείου for land destroyed and scoured away by the breach of a dike, 4 *arourai,* 2 *solidi.*
—*P.Oxy.* LVIII 3960.36–39 (621) reveals that the inhabitants of this locale were 12 *solidi* (on the private standard) in arrears; this sum had been converted to 600 *knidia* (i.e., 1 *sol.*/50 *knid.*), of which 7.333 *solidi* (or, more probably, their equivalent in wine) were given to a tribune for distribution to the Sasanians. The site may have already supplied 1,000 *knidia;* cf. line 11 and note.
—A *chōrion* Kotyleeiou appears on an undated potsherd (probably from a wine jar, cf. p. 43 above); see *SB* I 1973.16 (+ *BL* VIII 306). (The *epoikion* Kotyleeiou is only attested in the Byzantine and Islamic periods.)

* **ΚΡΑΡΙΟΥ** (*epoikion;* Lower Toparchy, eighth *pagus*)
—*PSI* VIII 954.22 (VI) reads, παρὰ τοῦ κοινοῦ τῶν γεωργῶν καὶ ἀμπελουργῶν ἐποικίου Σκυταλίτιδος καὶ Κραρίου ὑπὲρ ἀποτάκτου χωρίων, 52.5 *solidi.*

* **ΚΤΗΣΕΩΣ** (*epoikion, chōrion;* Western Toparchy, third *pagus,* just north of the city of Oxyrhynchus)[33]
—The *chōrion* Kteseos appears on an undated potsherd (*SB* I 1973.10), probably from a wine jar (cf. p. 43 above), and the *epoikion* Kteseos is known to have belonged to the Apiones (cf. *SB* XXII 15364 [582]). (There are no attestations of an Oxyrhynchite site with this name before the fourth century.)

ΛΕΟΝΤΟΣ (*epoikion, ktēma*)
—In *P.Oxy.* XVI 1896 (577), eight *enapographoi geōrgoi* from this *epoikion* (headed by a *phrontistēs*) agree to supply three thousand 8-*sextarius sēkōmata* plus the last indiction's arrears from the *geouchikai ampeloi* of their *ktēma.*
—In *P.Oxy.* XVI 2044.4 (probably 579), the *oinocheiristēs* Phib segregates 2,243 *dipla* of vinegar that have come from the *ktēma* Leontos.

33. It is described as περὶ Παεῖμι‹ν› κτήματος in *SB* XXII 15364.13.

—This *epoikion* appears to have belonged to the *geouchos* Tatianos in the fifth century; see *P.IFAO* II 12.[34] For the identity of this landowner, see *P.Oxy.* LXVIII 4680, 1n. (where it is stated, incorrectly, that the provenance of the IFAO papyrus is unknown), and 4685 back, 5n.

ΛΕΥΚΙΟΥ (*kōmē, chōrion;* Western Toparchy, third *pagus*)

—At first glance, *P.Oxy.* XVIII 2197 (ca. 585?; cf. *P.Oxy.* LXX 4792, 10n.) might seem to indicate that this locale was owned by the Apiones (cf. Mazza 2001, 182, s.v.), but the word following *epoikion* in the heading (at l. 86) is actually Loukiou; cf. *BL* IX 194. (I read Λ[ο]υ̣κ̣ί[ου] on a digital image.) With the exception of *P.Mich.* XIII 674, which has *chōrion*, all references in the papyri are to the *kōmē* Leukiou. The Michigan piece must date to the period after the Islamic conquest (though the ed.'s sixth-century date, based on the hand, seems appropriate) or refer to a vineyard in the village. Note also *SB* I 1945.12, a piece of Byzantine pottery, likely a wine vessel, reading *chōr( )* or *mer( )* Leuk( ).

ΛΕΩΝΙΔΟΥ (*epoikion, ktēma;* Middle Toparchy, seventh *pagus;* often yoked with ΛΙΜΕΝΙΑΔΟΣ)[35]

—*P.Oxy.* XVI 1910.12 (VI/VII) is a concession τοῖς ἀπὸ Λεωνίδου ὑπὲρ τῶν χωρίων κατὰ τὸ ἔθος, 40 *cancellus artabai.* This customary concession probably existed because the (grain) land had been converted to vineyard. In the following line, a concession is given for this very reason (cf. Hickey 1999, 325–26 = *BL* XII 142).

—In *P.Oxy.* XVI 2044.9–10 (probably 579), the *oinocheiristēs* Phib segregates 2,200 *dipla* of vinegar that have come from the *ktēmata* Leonidou *kai* Limeniados.

—*P.Oxy.* LVIII 3960.4 (621) records delivery of 6,200 *knidia* of wine from those *apo* Leonidou *kai* Limeniados; a second hand notes that this quantity is out of 7,822 *knidia,* but the arrears (l. 31) are listed only as 1,166 *knidia.*

ΛΙΜΕΝΙΑΔΟΣ (*ktēma;* Middle Toparchy, seventh *pagus*)

See ΛΕΩΝΙΔΟΥ.

ΛΟΥΚΙΟΥ(*epoikion, ktēma;* Upper Toparchy)[36]

—*P.Oxy.* XVI 1911.118 (557) is a concession to an *ampelourgos* for land ἐν αὐτουργίᾳ δικαίου τοῦ κτήματος Λουκίου (land taken for direct

34. Kruit and Worp redate this text to the fourth century (cf. *BL* XII 92) for unspecified reasons.
35. Cf. *PSI* I 52.9 (602/617/647; see Worp 2005, 155–58), where it is described as περὶ Τάμπετι.
36. Cf. *P.Oxy.* LV 3804, intr., pp. 96–97.

cultivation), .75 *solidus* (out of 2.156 *sol.*) for 1 *aroura*. This *ampelourgos* figures in other entries in 1911 (e.g., l. 128) and in the next item.

—*P.Oxy.* LV 3804.75 (566) is under the geographic heading for this site and has a sum due (2.531 *sol.*) from an *ampelourgos*.[37]

—*P.Oxy.* LV 3804.96 has a sum due παρὰ τοῦ κοινοῦ τῶν γεωργῶν ὑπὲρ ἐκτάκτων τρύγης,[38] .416 *solidus* + 400 myriads of *denarii*. Cf. the identical entry in *SB* XXIV 16324.25 (557).

—*P.Oxy.* LV 3804.101 reads, παρὰ τοῦ κοινοῦ τῶν γεωργῶν καὶ ἀμπελουργῶν ὑπὲρ ἀποτάκτου χωρίων, 9 *solidi*. Cf. the identical entry in *SB* XXIV 16324.31 (557).

—*P.Oxy.* LV 3805.87–88 (after 566) contains the following entry: διὰ τῶν ἀπὸ Ταρουσέβτ καὶ Λουκίου τῶν κτημάτων ὑπὲρ ἀποτάκτου χωρίων, 15 *solidi* less 75 *keratia* on the private standard, which is converted to 10 *solidi* + 16.75 *keratia* on the Alexandrian standard.

ΜΑΤΡΕΟΥ (*epoikion, ktēma;* Eastern, Western, or Upper Toparchy; third or fourth *pagus*)[39]

—This *epoikion* is associated with a *mēchanē* suggestively named Τῶν Χωρίων (*P.Oxy.* XIX 2244.31 [528/543/558; cf. Gonis 2004a, 201]). The settlement is attested as Apion property in, e.g., *P.Oxy.* I 136 (583).

* **ΜΗΤΡΟΔΩΡΟΥ**(*epoikion, ktēma*)

—*P.Oxy.* LXX 4781 (525) is a receipt for an axle for a *geouchikē mēchanē* that is said (l. 10) to supply water εἰς ἄμπελον καὶ εἰς ἀρόσιμον γῆν.[40] The recipient's *origo* is the *epoikion* Metrodorou.

* **ΜΟΥΧΕΩΣ, ΜΕΓΑΛΟΥ** (*epoikion;* Lower Toparchy)[41]

—*PSI* VIII 954.42 (VI) reads, παρὰ τοῦ κοινοῦ τῶν γεωργῶν καὶ ἀμπελουργῶν Μεγάλου [ed. pr.: Μεγάλ(ης)] Μούχεως τῆς πεδιάδος, 27 *solidi*.

37. This same *ampelourgos* and partners rent an oil factory (2.5 *sol.*, l. 77), and he is part of a group that is charged an unspecified 8 *solidi* (l. 87). The latter entry might correspond to a concession of 4 *solidi* for 6 *arourai* (out of 12, l. 205), while the *ampelourgos* alone receives a concession of 1.5 *solidi* for 2 *arourai* (l. 209). The lands connected with the concessions are termed ἐν αὐτουργίᾳ; i.e., they have been taken for direct cultivation.

38. For these supplements, see p. 69 above.

39. *P.Oxy.* I 136 (= *W.Chr.* 383; 583), e.g., reveals that it was in a *prostasia* with the *kōmai* Adaiou *kai* Episemou, so it was presumably in the vicinity of these villages.

40. See n. 12.

41. Because it appears in a *pronoētēs*' account (*PSI* VIII 954) with sites known to have been in this area. For Μεγάλου, as opposed to the expected Μεγάλης, see Pruneti 1981, s.v. ΜΟΥΧΕΩΣ.

—*P.Mich.* XV 734 (572) is a sale of wine in advance in which the seller-debtors are from this *epoikion;* there is no obvious connection, however, to the Apiones. *SB* XXII 15491 (V) may be relevant to the question of ownership.

ΝΕΚΩΝΘΕΩΣ (*epoikion, ktēma;* Lower Toparchy)[42]
Note: This site is associated with the *dikaion* Kephala in, e.g., *P.Oxy.* XVIII 2195.160–61 (576/77), but there is also a geographical heading in the same text that does not make reference to the *dikaion* (l. 20).[43]
—*P.Oxy.* VI 998.5 (VI) is a concession (22 *sol.* on the Alexandrian standard + 12 *ker.*) τοῖς ἀπὸ Νεκώνθεως ὑπὲρ ἀποτάκτου χωρίων.[44]
—*P.Oxy.* LV 3805.76–78 (after 566) contains the following entry: διὰ τῶν ἀπὸ Νεκώνθεως ὑπὲρ ἀποτάκτου χωρίων, 30 *solidi* less 120 *keratia* on the private standard, which is converted to 22 *solidi* + 15.75 *keratia* on the Alexandrian standard; cf. the preceding item.
—In Text I, there are four vineyards (5, 2, 2, 1.5 *ar.*), each rented at 150 *dipla/aroura* (total 1,575 *d*).

* **ΝΕΟΥ** (*epoikion, ktēma*)
—*P.Oxy.* XVI 1988 (587; complete edition: *P.Oxy.* LXX, pp. 137–38) is a receipt for an axle for a *geouchikē mēchanē* called Τῶν Χωρίων that furnishes water εἰς τὰ αὐτὰ χωρία.[45] The recipient's *origo* is the *epoikion* Neou.
—In *P.Oxy.* XVI 2044.5 (probably 579), the *oinocheiristēs* Phib segregates 1,700 *dipla* of vinegar that have come from the *ktēma* Neou.

* **ΝΗΣΟΥ ΛΕΥΚΑΔΙΟΥ** (*epoikion, ktēma;* Upper Toparchy)[46]
—*P.Princ.* II 88 (V) is a short list of arrears of wine (in *dipla;* cf. *BL* IX 220) from this *epoikion,* which is known to have belonged to the Apiones (cf. *P.Oxy.* I 134 [569]).

* **ΝΙΚΑΡΩΝΟΣ** (*epoikion, chōrion*)
—Attested as Apion property in *P.Oxy.* XVI 2025.29 (VI/VII). There is a potsherd from the Byzantine period, probably from a wine jar, that bears the words *chōrion* Nikar( ); see *SB* I 1973.8, and cf. p. 43 above.

42. Because it appears in a *pronoētēs'* account (*P.Oxy.* XVIII 2195) with sites known to have been in this area.
43. For *dikaion,* see pp. 58–60 above.
44. The text is a *descriptum* and will be published by Benaissa and the author elsewhere.
45. See p. 42 n. 21 above.
46. Cf. *P.Oxy.* LV 3804, intr., pp. 96–97.

* **ΝΙΚΗΣ** (*epoikion, ktēma, chōrion*)
—Attested as Apion property in, e.g., *SB* XII 11231 (549). There is a potsherd from the Byzantine period, probably from a wine jar, bearing the name of a *chōrion* Nikes; see *SB* I 1945.16, and cf. p. 43 above.

* **ΝΙΚΗΤΟΥ** (*epoikion;* Middle or Thmoisepho Toparchy)[47]
—*P.Oxy.* LXX 4788 (566) is a receipt for an axle for a *geouchikē mēchanē* that is said (ll. 12–13) to supply water εἰς ἄμπελον καὶ εἰς ἀρόσιμην γῆν.[48] The recipient's *origo* is the *epoikion* Niketou.

* **ΝΟΜΟΓΡΑΦΟΥ** (*epoikion, chōrion;* Middle or Thmoisepho Toparchy?)[49]
—Perhaps an Apion property; cf. *P.Oxy.* XVI 1912 (ca. 555; cf. *BL* X 144). There is a potsherd from the Byzantine period, probably from a wine jar, bearing the name of a *chōrion* Nomographou; see *SB* I 1945.17, and cf. p. 43 above. See also the undated potsherd *SB* I 1973.22.

* **ΟΛΥΜΠΙΑΔΟΣ** (*epoikion, chōrion*; first *pagus*)[50]
—Attested as Apion property in *P.Oxy.* XVI 2025.18 (VI/VII), it was once part of the *domus divina* (*P.Köln* XI 459 [436?]). There is an undated potsherd, perhaps from a wine jar, reading *chōrion* Olympi( ); see *SB* I 1973.11, and cf. p. 43 above.

* **ΟΡΘΩΝΙΟΥ** (*epoikion, ktēma, chōrion;* Western or Eastern Toparchy?)[51]
—In *P.Oxy.* XVI 2044.8 (probably 579), the *oinocheiristēs* Phib segregates 100 *dipla* of vinegar that have come from the *ktēma* Orthoniou.
—*P.Oxy.* LVIII 3960.35 (621) lists 514 *knidia* as arrears due from οἱ ἀπὸ Ὀρθωνίου. This site may have already supplied 2,000 *knidia;* cf. line 10 and note.
—*SB* I 1945.19 is a piece of Byzantine pottery reading *meris* or, more probably, *chōrion* Orth( ).

* **ΟΣΤΡΑΚΙΝΟΥ** (*epoikion;* Herakleopolite, Koites?)[52]
—*P.Oxy.* XVI 1917.80 (probably 616/17; cf. *BL* X 144) lists arrears (1.25 *sol.*) due from *ampelourgoi* in the *phrontis epoikiou* Ostrakinou.

47. It is yoked with the *epoikia* Terytheos *kai* Theagenous in *P.Oxy.* VI 998.11; cf. also *P.Oxy.* XIX 2244.59 and 63, which imply proximity to the *kōmē* Tampeti.

48. See n. 12.

49. This location if its presence in *P.Oxy.* XVI 1912.76 indicates proximity to the *prostasia* that is documented by this account.

50. Cf. *P.Oxy.* LXII 4342/1 (IV).

51. This location if proximate to the other sites mentioned in *P.Oxy.* XVI 2018.19ff. (VI). Benaissa (2009, s.v.) suggests the former Lower Toparchy.

52. Cf. Falivene 1997, 154.

ΟΥΑΛΕΝΤΟΣ (Middle or Thmoisepho Toparchy)[53]
—*P.Oxy.* XVI 1912.71 (ca. 555; cf. *BL* X 144) reads, παρὰ τῶν ἀπὸ Οὐάλεντος ὑπὲρ ἀποτάκτου χωρίων, 3 *solidi*,[54] while line 136 is a concession (amount lost) for the same.

ΟΥΕΣΩΒΘΕΩΣ (*epoikion, ktēma;* Middle or Thmoisepho Toparchy)[55]
—*P.Oxy.* X 1326 (V/VI) is a receipt (for 275 *d*) that is addressed to *ampelourgoi* of this *epoikion* from the *oinocheiristēs* Philoxenos.
—*P.Oxy.* XVI 1912.127–28 (ca. 555; cf. *BL* X 144) indicates that inhabitants of this site received empty jars for the "bottling" of wine.

ΠΑΓΓΟΥΛΕΕΙΟΥ (*epoikion, ktēma;* Western Toparchy, third *pagus*)[56]
—*P.Oxy.* XXXVI 2779 (530) is a receipt for an axle for a *geouchikē mēchanē* that is said (l. 11) to supply water εἰς ἄμπελον καὶ εἰς ἀρρόσιμον [read ἀρόσιμον] γῆν.[57] The recipient's *origo* is the *epoikion* Pangouleeiou.
—*P.Oxy.* LXI 4132 (619) is a sale in advance (8 *sol.* on the private standard for eight hundred 5-*sext.* measures plus "expenses of the vintage"); the seller-debtor is known to have been an *enapographos geōrgos* of the Apiones,[58] and the guarantor for the transaction is an Apion *pronoētēs.*

* **ΠΑΚΕΡΚΗ/-Υ**(*ktēma, kōmē;* Eastern or Middle Toparchy)[59]
—*P.Oxy.* LV 3805.68–69 (after 566) reads, διὰ τῶν κληρονόμων Ἰερημίου μείζονος Πακέρκυ ὑπὲρ ἀποτάκτου οἴνου, 12 *solidi* less 48 *keratia* on the private standard, which is converted to 9 *solidi* + 1.5 *keratia* on the Alexandrian standard.

ΠΑΚΙΑΚ (*epoikion, ktēma;* Upper Toparchy)[60]
—*P.Oxy.* LV 3804.47 (566) reads, παρὰ τοῦ κοινοῦ τῶν γεωργῶν καὶ ἀμπελουργῶν ὑπὲρ ἀποτάκτου χωρίων, 13.5 *solidi.*

---

53. See n. 13.
54. For the reading, see *BL* XI 155.
55. See n. 13.
56. Location contra Gomaà et al. 1991, 81; cf. *P.Oxy.* LXI 4132, 9–10n.
57. See n. 12.
58. See LXI 4132, 8–10n.
59. *Ktēma* and *kōmē* are not incompatible in this case, for at least two sites are known to have had this name; cf. Mazza 2001, 183 and refs. The entry here clearly refers to a village (note *meizōn*, "village headman").
60. Cf. *P.Oxy.* LV 3804, intr., pp. 96–97.

* **ΠΑΛΩΣΙΣ** (*ktēma;* Thmoisepho Toparchy; eighth *pagus*)
—The *origo* of the seller-debtor of 168 *chymata* of wine in the advanced sale *T.Varie* 8 (669), it was an Apion property before the Islamic conquest (cf. *P.Oxy.* LXX 4794 [580]).

ΠΑΡΘΕΝΙΑΔΟΣ (*epoikion, ktēma;* Western Toparchy)[61]
—*P.Oxy.* XVI 1990 (591; complete edition: *P.Oxy.* LXX, pp. 140–42) is a receipt for an axle for a *geouchikē mēchanē* that is said (ll. 18–19) to supply water εἰς ἄμπελον καὶ εἰς ἀρρώσιμον [read ἀρόσιμον] γῆν.[62] The recipient's *origo* is the *ktēma* Partheniados.
—In *P.Oxy.* XVI 2044.7 (probably 579), the *oinocheiristēs* Phib segregates 150 *dipla* of vinegar that have come from the *ktēma* Partheniados.
—In *P.Oxy.* XVIII 2197 (ca. 585?; cf. *P.Oxy.* LXX 4792, 10n.), under an account heading for the *epoikia* Partheniados *kai* Hiereon, bricks are expended for repairs on the house of an *ampelourgos* (l. 75).
—*P.Oxy.* LVIII 3960.34 (621) lists 153 *knidia* as arrears due from οἱ ἀπὸ Παρθενιάδος.

* **ΠΑΡΟΡΙΟΥ, ΜΕΓΑΛΗΣ** (*epoikion, ktēma;* Western Toparchy?)[63]
—*SB* XXII 15320 (2nd half V) is a receipt (for 219 *d*) that is addressed to an *ampelourgos* of this *epoikion* from the *oinocheiristēs* Philoxenos.
—*P.Oxy.* XVI 1913.33–35 (555?) reveals that it was also a source for wine jars.
—In *P.Oxy.* LXXV 5069 (546), an advanced sale of 23 8-*sextarius* measures of wine, the seller-debtor is from this locale.
—A *chōrion* Paroriou appears on an undated potsherd (perhaps part of a wine jar); see *SB* I 1973.4. There were several sites, however, whose names included this word. See also the Byzantine sherd *SB* I 1945.21.

* **ΠΑΡΟΡΙΟΥ, ΜΙΚΡΑΣ** (*epoikion;* Lower Toparchy)[64]
—*PSI* VIII 954.40 (VI) has a sum due παρὰ τοῦ κοινοῦ τῶν γεωργῶν [±8 = (καὶ) ἀμπελουργ(ῶν)?] ὑπὲρ ἐκτάκτων [ed. pr.: ἐκτάκ(της)] τρύγης, 2+ *cancellus artabai* and 6.375 *solidi* (sum read on original). This entry may, in fact, concern Psempekle.
—A *chōrion* Paroriou appears on an undated potsherd (perhaps part of a wine jar); see *SB* I 1973.4. There were several sites, however, whose names included this word. See also the Byzantine sherd *SB* I 1945.21.

---

61. *P.Oxy.* LXVIII 4704 implies proximity to the city of Oxyrhynchus; also see ΙΕΡΕΩΝ above.
62. See n. 12.
63. See n. 26.
64. See n. 41.

ΠΑΨΑΥ (*epoikion;* Middle or Thmoisepho Toparchy)[65]

—*P.Oxy.* XVI 1912.29 (ca. 555; cf. *BL* X 144) reads, παρὰ τῶν αὐτῶν γεωργῶν καὶ ἀμπελουργῶν, 1 *metron artabē,* 12.083 *solidi,* and 2,000 *denarii.*

—*P.Oxy.* XVI 1912.127–28 indicates that inhabitants of this site received empty jars for the "bottling" of wine.

—*P.Oxy.* XVI 1912.152–55 records the total price (11 *sol.* less 49.5 *ker.;* cf. table 3.2.2) of new reeds bought for newly planted *chōria* (2 *ar.* at this site).

—*P.Oxy.* XVI 1982 (497) is a receipt for an axle for a *geouchikē mēchanē* that is said (l. 11) to supply water εἰς ἄμπελον καὶ εἰς ἀρόσιμον γῆν.[66] The recipient's *origo* is the *epoikion* Papsaou.

* **ΠΕΜΠΩ** (*ktēma;* Lower Toparchy?)[67]

—This imperial *ktēma* was administered by the Apiones.[68] It included 4 *arourai* of vineland that were charged with *apotakton chōriōn* [ed. pr.: χωρ(ίου)], but the corresponding sum due was customarily reduced by 2 *solidi* ὑπὲρ ἐνχόρτου (see *P.Oxy.* XVI 1915.3–7 [post-557]).

ΠΕΡΑ (*ktēma*)

—In *P.Oxy.* XVI 2044.6 (probably 579), the *oinocheiristēs* Phib segregates 250 *dipla* of vinegar that have come from the *ktēma* Pera.

* **ΠΕΡΟΥΕΝ** (*epoikion, ktēma;* Lower Toparchy)[69]

—P.Oxy. XVI 2038 (VI–VII) is an account of remissions; under the heading for the *epoikion* Perouen, 23.75 *cancellus artabai* are conceded to a *koinon* of *ampelourgoi* ὑπὲρ τῆς ἐξωτικῆς γῆς ὑπὲρ ἀβρόχου (l. 20). The land in question was not vineland; presumably the *koinon* worked vineyards elsewhere. Cf. also *PapCongr. XXIV*:417–19.

ΠΕΣΤΑ (*epoikion, ktēma;* Western/Eastern/Middle Toparchy)[70]

—*P.Oxy.* XIX 2243a.29 (590) reads, παρὰ τῶν αὐτῶν καὶ προκειμένων γεωργῶν ὑπὲρ ἀποτάκτου χωρίων [ed. pr.: χωρ(ίου)], 7.5 *solidi* without *rhopē.* The preceding line is an entry for the price of a pig (.646 *sol.*) and

65. See n. 13.

66. See n. 12.

67. *P.Oxy.* XVI 1915 suggests that it was near Meskanounis, which Benaissa (2009, s.v.) indicates may have been in the former Lower Toparchy.

68. See discussion above, pp. 52–53.

69. It was in the same *prostasia* as Spania (cf. *P.Oxy.* XVIII 2204).

70. It appears in a *pronoētēs'* account (*P.Oxy.* XIX 2243a) with sites from these former toparchies; cf. Benaissa 2009, s.v.

concerns the same individuals; it could represent one of the expenses of the vintage or a vineyard rent supplement.

* **ΠΕΤΝΗ** (*chōrion, kōmē;* Middle or Thmoisepho Toparchy,[71] Upper Toparchy[72])

—A *chōrion* Petne appears in *P.Oxy.* XVI 2043 (V). There were two villages with this name, both of which may have been under Apion jurisdiction (cf. *P.Oxy.* XVI 1910, 1912, with Benaissa 2009, s.v.).

* **ΠΕΤΡΩΝΙΟΥ** (*chōrion*)

—A *chōrion* Petroniou appears in *P.Oxy.* XVI 2043 (V), where it is subdivided into three *merides.* A site with this name is linked to the Apiones; cf. *P.Oxy.* XVI 2031.5 (VI/VII), 2032.35 (540–41; so *BL* VI 105).

* **ΠΗΛΕΑΣ** (*ktēma*)

—Not a locale with Apion vineyards, but a source of wine purchased by the *oikos* (*P.Oxy.* LVIII 3960.13 [621], 450 *knid.* at 18 *knid./sol.*).

* **ΠΙΑΑ** (*epoikion;* Middle or Thmoisepho Toparchy)[73]

—*P.Oxy.* XVI 1899 (476; cf. *CSBE*^{2}, 89) is a receipt for a *kyklas* for a *geouchikē mēchanē* that is said (l. 11) to supply water εἰς ἄμπελον καὶ εἰς ἀρόσιμον γῆν.[74] Here the machine belongs to a certain Flavius Alexandros, but the site later appears in Apion documents.[75]

* **ΠΟΣΟΜΠΟΥΣ/ΠΟΥΣΕΜΠΟΥΣ** (*kōmē;* Eastern Toparchy, fifth *pagus*)

—A source of wine purchased by the *oikos* (*P.Oxy.* LVIII 3960.15ff. [621], 800 *knid.* at an unspecified price; 117 *knid.* were returned because they were vinegar).[76]

ΠΡΟΑΣΤΙΟΥ (*chōrion;* near the city of Oxyrhynchus; refers to the *proast(e)ion Exo tes pyles;* for the prepositional phrase alone, see above.)

—*P.Oxy.* XVI 1913.33–35 (555?) records that the *ampelourgoi* of the *chōrion proastiou* were the recipients of 764 new empty jars for the "bottling" of wine. The *chōrion proastiou* also appears in *P.Princ.* III 158.4 (VI).

71. This location if in the *prostasia* documented by *P.Oxy.* XVI 1912; see n. 13.

72. See Benaissa 2009, s.v.

73. It appears in the *prostasia* documented by *P.Oxy.* XVI 1912; see n. 13.

74. For *kyklas,* see Bonneau 1993, 112–13.

75. *P.Oxy.* XVI 1912 (ca. 555; see *BL* X 144), *P.Oxy.* XVIII 2207 (VI), and *P.Wash.Univ.* II 84 (late VI) suggest Apion ownership, while the entry in the margin of *P.Oxy.* XIX 2244 (528/543/558; cf. Gonis 2004a, 201) may indicate that it was at one time a possession of the church. Also see n. 12.

76. The editor mistranslates *oxos* as "new wine."

* **ΠΤΕΛ** (*epoikion*)
—Property of the Empress Eudocia administered by the first Strategius (*P.Oxy.* L 3585), it appears in *P.Oxy.* XVI 2042 (V) as the source of some wine (perhaps 70+ *knid.*) carried by *libernos* (liburnian). (Also note the ecclesiastical connection revealed by *SB* XII 10808 [VII, postconquest; see *CPR* XXII, p. 283].)

* **ΠΤΟΛΕΜΑΙΔΟΣ** (*epoikion, ktēma;* Lower Toparchy)[77]
—*PSI* VIII 954.48 (VI) reads, παρὰ τοῦ κοινοῦ τῶν γεωργῶν ὑπὲρ ἀποτάκτου χωρίων, number of *solidi* not extant.

ΣΑΜΑΚΙΩΝΟΣ (*epoikion;* Middle or Thmoisepho Toparchy)[78]
—*P.Oxy.* XVI 1912 (ca. 555; cf. *BL* X 144) has several entries with amounts due from *ampelourgoi* and *geōrgoi* of this site (l. 85: 3 *metron art.*, 6.771 *sol.;* l. 86: more than .199 *sol.;* l. 88: 8 *sol.* [probably *apotakton chōriōn;* cf. *BL* XI 155]).

ΣΕΡΥΦΙΣ (*kōmē;* Western Toparchy, third *pagus*)
—*P.Oxy.* XVI 1859 (VI/VII) is a request from a *chartularius* that Pambechios and his partners, who are from Seryphis, be allowed to finish their work on the vintage.

ΣΚΥΤΑΛΙΤΙΔΟΣ (*epoikion;* eighth *pagus*)
—*PSI* VIII 954.7 (VI) has a sum due παρὰ τῶν κληρονόμων τῶν γεωργῶν ἐποικίου Σκυταλίτιδος ὑπὲρ ἐκτάκτων [ed. pr.: ἐκτάκ(της)] τρύγης, 6 *cancellus artabai*, 4.875 *solidi*, 7,800 myriads of *denarii.*
—*PSI* VIII 954.22 reads, παρὰ τοῦ κοινοῦ τῶν γεωργῶν καὶ ἀμπελουργῶν ἐποικίου Σκυταλίτιδος καὶ Κραρίου ὑπὲρ ἀποτάκτου χωρίων, 52.5 *solidi.*
—*P.Oxy.* LXX 4799 (586) is a receipt for a cogwheel for a *geouchikē mēchanē* that is said (ll. 13–14) to supply water εἰς ἄμπελον καὶ εἰς ἀρρόσιμον [read ἀρόσιμον] γῆν.[79] The recipient's *origo* is the *epoikion* Skytalitidos.
—*P.Col.* XI 303 (515; new edition = Papathomas 1999 = *SB* XXVI 16753) suggests that this was not an Apion holding at the beginning of the sixth century.

77. See n. 41.
78. See n. 13.
79. See n. 12.

ΤΑ(Α)ΜΠΕΜΟΥ (*kōmē;* Eastern Toparchy, fifth *pagus*)

—*P.Oxy.* XIX 2243a.47–48 (590) has the following entry: παρὰ τοῦ κοινοῦ τῶν γεωργῶν ἀπὸ τῆς αὐτῆς ὑπὲρ φόρου τῶν γηδίων φυτευθέντων ἐν ἀμπέλῳ ἐν μηχανῇ Τεπότ, 24 *solidi* less 108 *keratia*, with *rhopē.*[80]

—*P.Oxy.* XIX 2244.86–88 (528/543/558; cf. Gonis 2004a, 201) indicates that two axles were supplied for two *kykleutēria* of a new reservoir (*lakkos*) that was being built εἰς τὰ νεόφυτα χωρία ἐν κώμῃ Ταμπέμου.[81]

* **ΤΑΜΠΕΤΙ** (*kōmē;* Middle Toparchy, seventh *pagus*)

—*P.Oxy.* XIX 2244.81–85 (528/543/558; cf. Gonis 2004a, 201) indicates that axles were supplied to a headman of this village for irrigation machinery in a certain *edaphos.* In both cases, the machinery is described as supplying water εἰς τὰ νεόφυτα χωρία διὰ τῶν ἀπὸ Ταμπετί. Though it is clear that Tampeti was administered in some capacity by the Apiones and included Apion holdings (cf. next entry), the axles are said to be ἐκ τῶν τῆς ἐκκλησίας, a possible indication that the church possessed the vineland in question. At a minimum, the machinery fell within the church's share of the *munus* documented by 2244, and it seems unlikely that this would have been the case if the Apiones owned it.

—*P.Oxy.* LXX 4785 (530) is a receipt for an axle for a *mēchanē* that is said (l. 14) to supply water εἰς ἄμπελον καὶ εἰς ἀρόσιμον γῆν. The recipient's *origo* is the *kōmē* Tampeti, which is described (ll. 11–12) as *pagarchoumenē kai* (ed.: "administered *also* [my emphasis]") by Strategius II, a description that seems compatible with the church presence posited in the preceding entry. The *mēchanē,* however, is certainly Apion property, since it is described as *geouchikē.*

* **ΤΑΡΟΥΘΙΝΟΥ, ΜΕΓΑΛΗΣ** (*epoikion, ktēma;* Upper Toparchy?)[82]

—Clearly identified as an Apion property in *P.Oxy.* I 133 (= *W.Chr.* 384; 579), it is ubiquitous as an *origo* for seller-debtors in sales of wine in advance:

—*P.Col.* VIII 245 (VI): 10.3 *solidi* on the private standard—7 to be

80. I am presuming the referent for αὐτῆς is in the heading in line 35 (i.e., *kōmē* Tampemou) and is not the "Megale Ousia" in the preceding line.

81. For *kykleutēria,* see Bonneau 1993, 110–11.

82. Upper Toparchy: so Rea, *P.Oxy.* LV 3804, 164–67n., regarding the *epoikion* Tarouthinou (i.e., the toponym without *Megales*), and note *P.Oxy.* I 135 (= *W.Chr.* 384), lines 14 and 33, where the two places are apparently equated.

repaid before the vintage at 100 5-*sextarii* measures per *solidus,* the remainder at the rates current at the vintage.

—*P.Mich.* XV 743 (622; see Gonis 2004a, 198): The buyer-creditor is an Apion notary. 6 *solidi* less 2 *keratia* for 550 5-*sextarius* measures.

—*P.Wisc.* I 11 (VII) + *BL* VI 70: The buyer-creditor is a *tektōn.* 1 *solidus* of 22.5 *keratia* for 55 *knidia* of 4 *sextarii* each. (This text dates from after the Islamic conquest [more precisely, 646/661/676; cf. *BL* VIII 201 and Gonis 2004a, 198 n. 15] and thus cannot have any link to the Apiones.)

—*SB* V 8264 (524; cf. *BL* VII 198): The buyer-creditor is called *thaumasiōtatos.*[83] Five hundred measures.

—*SB* XIV 11617 (580; cf. *BL* VIII 373): The buyer-creditor is called *thaumasiōtatos.* 3 *solidi* on the private standard at the rate current in the *pedias* of the seller-debtor's *ktēma.*

—*P.Princ.* II 87 (612; cf. Ziegler 1995, 193) is a sale in advance of an unknown commodity (given its provenance, very probably wine).[84] The buyer-creditor is a *thyrouros* of Hagios Theodoros.

—A *ktēma* Tarouthinou (i.e., the toponym without *Megalēs,* but cf. n. 82) is known to have been a production site for wine jars; cf. *P.Oxy.* XVI 1911.184–85 + *BL* XI 154–55 (557).

—SB XXII 15271 (V) may indicate that Tarouthinou (the toponym without *Megalēs,* but cf. n. 82) was not always an Apion possession.

ΤΑΡΟΥΣΕΒ(Τ) (*epoikion, ktēma;* Upper Toparchy)[85]

—*P.Oxy.* LV 3804.102 (566) associates the *epoikion* Tarousebt with the *dikaion* Diogenous (cf. p. 59).

—*P.Oxy.* XVI 1911.175–77 (557) records the total price (2 *sol.* on the private standard, with *rhopē*) of new reeds bought for newly planted *chōria* (.25 *ar.* at this site).

—*P.Oxy.* XVI 1911.181ff. records that wine jars were furnished to the *ampelourgoi* of this locale;[86] cf. *P.Oxy.* LV 3804.218–19 (where they receive a portion of eleven hundred used wine jars).

—*P.Oxy.* LV 3804.104 has a sum due παρὰ τῶν αὐτῶν καὶ προκειμένων τῶν γεωργῶν ὑπὲρ ἐκτάκτων τρύγης, 5.083 *solidi;* cf. also *SB* XXIV 16324.34 (557).

---

83. For the significance of *thaumasiōtatos,* see p. 99 above.
84. Cf. Hickey 1999, 326 = *BL* XII 163.
85. Cf. *P.Oxy.* LV 3804, intr., pp. 96–97, and Gomaà et al. 1991, 92.
86. For corrections to these lines, see *BL* XI 154–55.

—*P.Oxy.* LV 3804.223–24 records a payment of wages to a *potamitēs* and his partners working on an *aroura* of vineland that had been destroyed by water.

—There was a *mēchanē* suggestively named Τῶν Χωρίων in this *epoikion* (*P.Oxy.* LV 3804.249, XVI 1911.158;[87] note also XIX 2244.79 [528/543/558; cf. Gonis 2004a, 201]).

—*P.Oxy.* LV 3805.87–88 (after 566) contains the following entry: διὰ τῶν ἀπὸ Ταρουσέβτ καὶ Λουκίου τῶν κτημάτων ὑπὲρ ἀποτάκτου χωρίων, 15 *solidi* less 75 *keratia* on the private standard, which is converted to 10 *solidi* + 16.75 *keratia* on the Alexandrian standard.

* **ΤΕΡΥΘΕΩΣ** (*epoikion* [in text cited below; *kōmē* elsewhere]; *kōmē*: Eastern Toparchy, fourth *pagus; epoikion:* Lower Toparchy?)

—*P.Oxy.* XVI 2019.19 (547–48/562–63; cf. *BL* XII 144) reads, παρὰ τῶν ἀμπελουργῶν τοῦ αὐτοῦ κτήματος, *solidi* > .031. It is not clear, however, if "the same *ktēma*" means Terythis, which is mentioned in line 15. ἀπὸ τοῦ αὐτοῦ occurs as early as line 6, with no geographic reference preceding except the name (ll. 3–4) of the *prostasia* οἴκ(ων) Τερύθεως καὶ Θεα̣γέν̣[ου]ς̣ κ̣[αὶ Εὐτ]υ̣χ̣ιάδος καὶ ἄλλ(ων) ἐξωτικ(ῶν) τόπων.[88]

ΤΙΛΛΩΝΟΣ (*epoikion, ktēma;* Lower Toparchy)[89]

—*P.Oxy.* XVIII 2195.17 (576/77) reads, παρὰ τοῦ κοινοῦ τῶν γεωργῶν ἐποικίου [. . . .]ος ὑπὲρ ἐκτάκτης [so ed.; actually abbreviated εκτακτ( ) = ἐκτάκτ(ων)] τρύγης, 3 (presumably *art.*) of grain, an uncertain number of *solidi,* and 2,700 *denarii.* The entry appears under the heading for the *epoikion* Tillonos, and though the editor questions such a restoration (17n.), I read [Τίλλω]ν̣ος at the lacuna.

—*P.Oxy.* XVIII 2195.124 is a concession of 1 *solidus* to those ἀπὸ Τίλλωνος ὑπὲρ ἀποτάκτου χωρίου [so ed.; actually abbreviated χωρ( ) = χωρ(ίων)] μηχανῆς Κολλούθου. See also line 18 (where the restoration [Τίλλωνο]ς is probable; cf. l. 17 and the discussion just above).

87. For the reading, see *BL* XI 154.

88. In 2019.3, προστ̣[άτο]υ clearly should be προστασίᾳ; cf. *P.Oxy.* XVIII 2204.5 (ca. 565–66), and see also Azzarello 2006, 227 (though the correction was already made in Hickey 2001). I do not know how οἴκ(ων) fits, but I have my doubts about it. Perhaps "Mikras" (cf. *P.Oxy.* XVI 1983.8) should have been read since the *village* Terythis is known to have been in the fourth *pagus,* while the *epoikion* Theagenous was in the eighth (cf. *P.Wisc.* I 12 + *BL* VIII 201), a distance probably too great for joint administration. The *epoikion* Mikras Terytheos Aetiou (1983) has not been localized. Azzarello has suggested reading ἐ]π|οικί(ων). *P.Oxy.* XVI 2019 was missing when I visited Cairo in 1997.

89. See n. 16.

TOE (*ktēma;* Lower Toparchy?)[90]

—*P.Oxy.* XVI 1913.51–53 (555?) indicates that wine jars were made here for the *ampelourgoi* of the *epoikion* Knapheon (cf. above); the site was also a source of *baukalia* for grapes (ll. 49–50; presumably not intended for wine).

ΦΑΓΚΟΝΑ (*epoikion;* Lower Toparchy?)[91]

—Text I (which may indicate that it was part of the *dikaion* Kephala) registers a rent of 516 *dipla.* The area of the associated vineland was at least 3 (but less than 4) *arourai;* a character (or characters) on the papyrus representing a fraction may have faded. If no fraction is missing, the fixed-rate rent at the *epoikion* Phankona was higher than that at the *epoikion* Nekontheos.[92]

—For *PSI* VIII 954.56–57 (VI), see p. 78 n. 94 above.

ΦΑΤΕΜΗ(Ν)Τ[93] (*epoikion;* Western Toparchy)[94]

—*P.Oxy.* LXIII 4391 (471) is an order from Flavia Isis to an *ampelourgos* from this site (perhaps for the distribution of 429 *dipla* that had been purchased in advance at 11 *sol.*).

—*SB* I 1969 + *BL* II.1 19 reveals that a certain *meris* here was sharecropped; the landlord's share was 334 *dipla* and 6 *laē.*

* **ΧΕΝΕΤΩΡΙΟΣ** (*epoikion;* Middle or Thmoisepho Toparchy)[95]

—*P.Oxy.* XVI 1912.58 (ca. 555; cf. *BL* X 144) reads, παρὰ τῶν γεωργῶν καὶ ἀμπελουργῶν, 4.333 *solidi.* Also note the concession for *apotakton chōriōn* in lines 137–39.

—*P.Oxy.* XVI 1912.152–55 records the total price (11 *sol.* less 49.5 *ker.;* cf. table 3.2.2) of new reeds bought for newly planted *chōria* (1.5 *ar.* at this site).

---

90. The Lower Toparchy if the presence of individuals from this site under the heading of the *prostasia* Nekontheos *dikaiou* Kephala (*P.Oxy.* XVIII 2195.160ff.) indicates proximity to that district. But a *kerameus* from Toe also appears in *P.Oxy.* XVI 1913, i.e., in a text concerning a *prostasia* that was closer to the city of Oxyrhynchus.

91. It is one of three sites with which Text I is concerned and was probably near the *epoikion* Nekontheos. Support for this would seem to be provided by its presence (in an entry among the *analōmata*) in *PSI* VIII 954, which concerns a *prostasia* in the former Lower Toparchy.

92. See p. 174 above.

93. For comments on the orthography of this place name, cf. *P.Oxy.* LXIII 4391, 1n.

94. *P.Oxy.* XVI 1913.67–68 indicates that individuals from the *epoikion* Phatement were involved in the irrigation of gardens and orchards in the Apion holdings just outside the city of Oxyrhynchus. Here it may be noted that *SB* XX 15167 (+ *BL* X 234) presumably belongs to the Apion dossier, for the prepositional phrase *syn rhopē* (ll. 2, 5) does not occur outside of the family's papers.

95. See n. 13.

* **ΧΡΥΣΟΧΟΟΥ** (*epoikion;* Lower Toparchy)[96]

—*PSI* VIII 954.29 (VI) reads, παρὰ τῶν αὐτῶν καὶ προκειμένων γεωργῶν ὑπὲρ ἀποτάκτου χωρίων, 27 *solidi.* It is not certain, however, that this entry falls under the geographic heading for this *epoikion.* The Apiones may have owned only a third of this *epoikion* (cf. 954.26).

ΧΩΡΙΟΥ, ΜΕΓΑΛΟΥ (*epoikion*)

—The name of this Apion *epoikion* (cf., e.g., *P.Oxy.* XVI 1915, 2032) is suggestive, but no concrete evidence for viticulture presently exists.

* **ΧΩΡΙΟΥ, ΝΟΤΙΝΟΥ** (*epoikion*)[97]

—In *P.Oxy.* LVIII 3960.9 (621), the inhabitants of this *epoikion* supply 1,236 *knidia* of wine; 40 *knidia* remain in arrears (l. 32). (The name of this *epoikion* is suggestive, but note *P.Oxy.* LXX 4797, 6–7n.)

* **ΨΕΜΠΕΚΛΗ** (*epoikion;* Lower Toparchy)[98]

—*PSI* VIII 954.41 (VI) reads, παρὰ τοῦ κοινοῦ τῶν γεωργῶν καὶ ἀμπελουργῶν ἐποικίου Ψεμπέκλη ὑπὲρ ἀποτάκτου χωρίων, 9 *solidi.* See also ΠΑΡΟΡΙΟΥ, ΜΙΚΡΑΣ.

* **]ος, an *epoikion*** (Lower Toparchy)[99]

—Under this geographical heading in *PSI* VIII 954.18 (VI), the following appears: παρὰ Ἀείωνος ἀμπελουργοῦ ἀπὸ τοῦ αὐτοῦ κτήματος, .177 *solidus.* (This small sum would seem unlikely to be a vineyard rent.)

* **]ụ, preceded by** καὶ τῶν ἐπọ[ικ(ίων (*epoikia;* Middle or Thmoisepho Toparchy)[100]

—*P.Oxy.* XVI 1912.81 (ca. 555, cf. *BL* X 144) reads, παρὰ τῶν αὐτῶν γεωργῶν καὶ ἀμπελουργῶν ὑπὲρ ἀποτάκτου χωρίων [ed. pr.: χωρ(ίου)], 18 *solidi.* This entry appears under the heading for these unknown sites.

Unknown

—*P.Oxy.* LXX 4782 (528) is a receipt for an axle for a *geouchikē mēchanē* that is said (l. 4) to supply water to orchard and, possibly, vineland (see 4n.).[101] The recipient's *origo* is unknown.

---

96. See n. 41.

97. Both Pruneti (1981, 124) and Krüger (1990, 283) take *chōrion* as a common noun and therefore confound this *epoikion* (note the explicit reference in *P.Oxy.* LVIII 3953.8, ἀπὸ ἐποικίου Νοτίνου Χωρίου) with the toponym to which reference is made, e.g., in *P.Oxy.* XIX 2443a.76, ἐκκλησίαν Νοτινοῦ (*sic*).

98. See n. 41.

99. See n. 41.

100. See n. 13.

101. See n. 12.

Appendix C

Wine Containers Used in the Byzantine Oxyrhynchite

Within only the last decade or so, there has been a flurry of scholarship on the subject of measures and containers in the papyri,[1] so it would be quite reasonable to question the need for this appendix, especially since the continued stream of text editions is still far from rendering these recent contributions obsolete; in truth, the number of new texts cited below is relatively small. What this recent scholarship has lacked in large part, however, is a sifting based on spatial (as opposed to temporal) criteria. The geographic granularity of the data is important because local practice is a critical element in nomenclature and standardization (cf. also, e.g., the "local standards" for gold coinage),[2] and these "micropatterns" (including frequencies) are much more difficult to discern when attestations have not been sorted by place.[3] Siftings

1. Much of it derives from the efforts of Klaas Worp, Nico Kruit, and Philip Mayerson; see the entries under these names in the Literature Cited.

2. By "local," I mean not only the nomes but smaller spatial units, including estates. Of course, reference to administrative boundaries must be heuristic, for as a rule, practice would not have been contained by them.

3. I do not deny that certain ceramic centers produced containers for sale beyond their immediate region; cf. Wickham 2005, 761–62, on the staggering mounds of L(ate) R(oman) A(mphora) 7 amphorae at Hermopolis and, especially, Antinoopolis. I will revisit this issue elsewhere; for the present purposes, it seems sufficient to note that the Oxyrhynchite has its own such mounds (cf., e.g., Ballet and Picon 1987, 38) and that papyri from the region do not attest imports of containers in significant quantities (on the contrary, cf. p. 73 above). Wickham himself acknowledges that Hermopolis and Antinoopolis were not the only sources for LRA 7 amphorae, and I suspect that my argument for regionalism finds support in the diversity of types within the LRA 7 group (cf., e.g., Tomber and Williams 2000, 45).

of this sort are also valuable because the spatial units of inquiry for the archaeological data—that is, the actual vessels—tend to be smaller, be they regions surveyed or sites excavated. The systematic integration of textual and archaeological data concerning the ceramics of Byzantine Egypt remains a desideratum (and an underexploited opportunity to write economic history).[4] Perforce, this will be a collaborative ("interdisciplinary") project, and for the Oxyrhynchite, which lacks publications comparable to Bailey 1998, much preliminary work remains on the archaeological side.[5] In the interim, I cannot help but put forward one hypothesis for further investigation: that the most commonly attested containers for wine in the late Oxyrhynchite, the *knidion* and *diploun*—often one and the same vessel (see entries below)—correspond to L(ate) R(oman) A(mphora) 7, the ubiquitous wine amphora of Byzantine (and Islamic) Egypt. Prima facie, capacity might seem to be a hurdle for this identification; the vessels from the LRA 7 family measured by Bailey held between 3.65 and 8.6 liters.[6] But these amphorae could also be smaller, as Egloff 174 was at 2.7 liters.[7] Taking the Egloff capacity as a minimum yields a range of 5–16 *sextarii*, a range that nearly includes all of the *knidion/diploun* volumes attested in late Oxyrhynchite papyri (i.e., 4, 4.5, 6, and 8 *sextarii*).[8] All of these Oxyrhynchite volumes, of course, could have fit within LRA 7 vessels.[9] Non-Oxyrhynchite vessels that are common and have identical capacities (e.g., the *angeion*) should also be considered for membership in the LRA 7 family.[10]

---

4. Mayerson 1992 is the only work cited in the guide below that makes connections between textual and material evidence.

5. For examples of Oxyrhynchite pottery, see Petrie et al. 1925, pl. 48, as well as the "online catalogue" of the Petrie Museum at http://petriecat.museums.ucl.ac.uk/.

6. Bailey 1998, 130.

7. Egloff 1977, 1:115. The differences in LRA 7 capacity are quite clear in fig. 2 of Tomber and Williams 2000.

8. But note p. 123 n. 129 above, where it is suggested that the smaller capacities correspond to wineskins.

9. Vessels would not have been filled to capacity; cf. *Geoponica* VII.6.10: δεῖ δὲ τὸν μεταγγιζόμενον εἰς τὰ κεράμια οἶνον, οὐ μέχρι τοῦ χείλους τῶν κεράμων ἐμβάλλεσθαι, ἀλλ᾽ ἕως ὑποκάτω μικρὸν τοῦ τραχήλου, ὥστε μὴ πνίγεσθαι, ἀλλὰ διάπνοιαν ἔχειν. (Note Mayerson 2001a, 219–20, and Vogt et al. 2002, 68–69, on holes in the "shoulders" of amphorae.) Concerning the filling of amphorae, Mayerson (2002, 205–6) cites *Geoponica* VI.12.2, but this passage refers to fermentation vats (*pithoi*): These are not to be filled too high, lest the must "boil over" during fermentation. While Mayerson (2002 and elsewhere) is right to note that fermenting must was carried away in amphorae by the buyer-creditor in certain sales in advance, he gives this too much emphasis; the typical locus for (primary) fermentation was the *pithos.* Regarding fermentation in amphorae, cf. also Vogt et al. 2002, 68–69.

10. Cf. Kruit and Worp 1999, 116, which needs more precision: these vessels must be related by more than just their capacities. But the authors are to be commended for revealing the geographical patterns of the nomenclature.

The guide to papyrological attestations of Oxyrhynchite wine containers that follows is arranged in (Greek) alphabetical order. After the citations themselves and references to pertinent secondary literature, I have included information on the capacities of each vessel (when available) and noted the commodities besides wine (*sensu stricto*) that were stored in them.[11] Texts cited may have more than a single attestation of the container or measure at issue. Attestations not mentioned under "other products" have the vessel either as a wine container or without any specification of contents. Containers that are unambiguously empty are identified as such. Though I have not set out here to document the situation throughout Byzantine Egypt, I have included relevant non-Oxyrhynchite material that is known to me.

ἀγγεῖον (*angeion*): A common container for wine in regions upstream from the Oxyrhynchite (i.e., the Hermopolite, Antinoopolite, and Antaiopolite), it is attested but once as such in the Oxyrhynchite documentation (*SB* XX 14644). Cf. also Kruit and Worp 2002b, 45–51 and refs.

Oxyrhynchite attestations: *P.Oxy.* X 1343 (VI), XVI 1837 (VI), 1924 (V–VI), 2053 (VI); *SB* XX 14644 (VI). (I exclude *P.Oxy.* XVI 2058.18 [VI; oil] because it is conjecture.)

capacities: Not specified in the Oxyrhynchite attestations; elsewhere 5–8 *sextarii.*

other products: radish oil (*P.Oxy.* XVI 1837.4–5, XVI 2053.20); soap (*P.Oxy.* X 1343.4; cf. *BL* VI 101); olive oil (*P.Oxy.* XVI 1924.7); empty (*P.Oxy.* XVI 1924.9).

notes: For *P.Oxy.* X 1290.6, 10 (V), see Russo 1999, 10. Note also the diminutive *angidion,* which is attested in *P.Oxy.* XVI 1923.20 (V–VI; oil).

Ἀσκαλώνιον (*Askalōnion*): Discussed by Mayerson 1992, who suggests that it is a smaller jar and identifies it with Killebrew B, a shorter and broader jar, 40–55 centimeters in height, that is approximately half the size of the Gaza jar (on which see below). See also Kruit and Worp 2000b, 99–101 and refs., as well as *Lexicon Vasorum Graecorum* 1:308–9.

11. The notion of commodity specificity appears to evaporate in the sources. Contrast Empereur 1993, 41: "The Greeks of Egypt recognized a shape very well, not needing a label or *dipinto* in the least: the form of the amphora was sufficient in itself; for them it had a value relevant for the nature of its contents (just like in our modern Western world, for that matter)." See also Van Alfen 1996, 209.

Oxyrhynchite attestations: *P.Oxy.* XVI 1924 (V–VI), LVI 3862 (IV–V), LIX 4005 (VI); *SB* XX 14625 (V–VI).
capacities:—
other products: beans (*P.Oxy.* LVI 3862.25); empty (*P.Oxy.* XVI 1924.10; in l. 3, they contain wine); *rosatum* (*SB* XXII 14625.19).
notes: Ascalonian wine is attested in *P.Oxy.* XI 1384 (= Mertens-Pack[3] 2410; V) and *SB* XXII 15712.2 (+ Sijpesteijn 1994). Note also the *chilia Askalōnia oinou* sent by St. John the Almoner as aid for the Jerusalem Patriarchate (*Vita Ioannis Eleemosynarii* 18.16–17, ed. Festugière and Rydén, p. 366).

βαύκαλις, βαυκάλιον (***baukalis, -ion***): "... a sort of pot with a narrow neck," its name being (supposedly) both Alexandrian and onomatopoeic; see *P.Oxy.* LIX 4002, 7n. For its use as a cooler, cf. *P.Benaki* 4 (= *SB* XXVI 16686; IV, Arsinoite?), 7n. (but in this Athens text, *baukalion* apparently is to be understood as a generic word for "jar"). Also a measure containing three thousand bricks (cf., e.g., *P.Oxy.* XVIII 2197, passim); a plausible explanation for this is given in the just-cited note to *P.Oxy.* 4002. *P.Wash.Univ.* II 105.1 has the only possible oenological usage in the Oxyrhynchite corpus; the context is quite suitable, but one cannot be certain. See also *Lexicon Vasorum Graecorum* 2:59–66.
Oxyrhynchite attestations (as a vessel): *P.Oxy.* XVI 1913 (555?), LIX 4002 (IV–V); *P.Wash.Univ.* II 105 (VI–VII).
capacities:—
other products: grapes (*P.Oxy.* XVI 1913.49).
notes: All texts of secure provenance with diminutive –ιον are from the Oxyrhynchite. (HGv gives the provenance of *SB* XXVI 16686 as "Arsinoites," but this is only the findspot.) *Lexicon Vasorum Graecorum* 2:60 notes that many of the (literary) attestations "riconducono ad un ambito prevalentemente monastico."

Γαζίτιον (***Gazition***): Discussed by Mayerson 1992, who identifies it as Killebrew A, a jar of 70–85 centimeters in height. Note also Peacock and Williams 1986, class 49, and Kruit and Worp 2000b, 100–103 with refs., as well as *Lexicon Vasorum Graecorum* 3:42–43. *P.Oxy.* XVI 1924 (V–VI) is the only secure Oxyrhynchite attestation; Gaza was a famous wine-producing region, and *P.Got.* 17 (new edition: *Eranos* 101 [2003] 114–20) indicates that the Gazition carried wine, but it appears in the aforementioned Oxyrhynchus papyrus only as an empty vessel (l. 11) and as a container for a pickled foodstuff (l. 9).

διδιπλοῦν (*didiploun*): Sijpesteijn (1982) suggests a measure of two *dipla*, but since it appears to be the equivalent of an *angeion* of 8 *sextarii* in *P.Köln* IV 192.6–7 (Antinoopolite, V–VI; note, however, the comments of Worp 2000, 148–49) and since a *diploun* of 8 *sextarii* is known (cf. next entry), I am inclined simply to translate as "a double double." (Note also *SB* XX 15202 [Antaiopolite, 1st half VI], where 140 *knid.* [l. 11] are equated with 70 *didipla* [l. 29 + *BL* X 234].) All attestations are Byzantine. In addition to wine (including must and *palaios oinos*), it is attested holding turnips (*CPR* IX 28.3 [Hermopolite, V–VI]). *P.Wash.Univ.* II 105.3 (VI–VII) is the only Oxyrhynchite attestation.

διπλοῦν (*diploun*): Cf. Worp 2000, 146–48 and refs., as well as *Lexicon Vasorum Graecorum* 4:156–62.

Oxyrhynchite attestations: *P.Harr.* I 100 (V); *P.Haun.* III 68 (402); *P.Laur.* II 42 (V + *BL* XII 97), IV 178 (2nd half VI), 185 (VII; misread as ψ(αθία) in ll. 14, 15); *P.Mich.* XIII 674 (VI? cf. p. 172 above); *P.Oxy.* I 141 (= *Stud.Pal.* VIII 1155; 503), VI 992 (413), 993 (478/79 or 493/94; cf. Gonis 2004a, 199–200), X 1322 (413), 1325 (V), 1326 (V/VI), XIV 1720 (VI; pace *BL* XII 141), XVI 1870 (V), 1893 (535), 1916v (551/52; = Text I), 1920 (563; cf. *BL* X 145), 1923 (V–VI), 1945 (517), 1950 (487), 1951 (485?; cf. Gonis 2004a, 199–200), 2043 (V), 2044 (probably 579), 2046 (563?; cf. p. 123 above), 2049 (VI), 2051r (VI–VII), XVIII 2196r (after 587; see p. 16 n. 103 above), XIX 2244 fr. 1 (after 570/71; cf. p. 90 n. 3 above), XXVII 2480 (580/81; cf. pp. 92, 95–97 above), XLIII 3148 (424), LI 3631 (V), LVI 3863 (V), 3868 (VI), LXIII 4391 (471), LXVII 4621 (497/512/527?; cf. 3n.), 4622 (V–VI), LXVIII 4683 (426), 4699 (504), LXXII 4901 (408); *P.Princ.* II 88 (V; see *BL* IX 220); *PSI* III 165 (441/42 + *BL* VIII 393), 191 (565), 192 (565), 193 (565), 194 (566?; so ed. pr.; cf. also *PSI* VI 709.6ff.), V 478 (V), VIII 920 (VI), 953 (576/68), 955 (ed.: VI; but see p. 91 n. 4 above), 956 (567/68), 957 (504 + *BL* XI 248–49); *P.Wash.Univ.* II 87 (V–VI), 98 (V), 99 (V–VI), 105 (VI–VII); *SB* I 1969 (VI), XVI 12608 (511), XVIII 13921 (507), XX 14625 (V–VI; see *BL* XII 227), 15181 (see *PSI* VIII 955 above), XXII 15320 (2nd half V), 15528–30 (497/512/527?; cf. *P.Oxy.* LXVII 4621, 3n.).

capacities: 4.5 *sextarii* (*P.Oxy.* XVI 2046.7, 15); 6 *sextarii* (*P.Oxy.* XVI 1920.7, 10, 12, 18, XXVII 2480.50–54, 57–60);[12] 8 *sextarii* (*P.Oxy.* XXVII 2480, *PSI* VIII 953, etc., with pp. 118, 124 above; no doubt the standard

12. The conversion from *sextarii* to *dipla* in *P.Oxy.* XVI 2046.57 is 4.5:1, but the remainder, .5 *sextarius*, is converted to .083 *diploun*, a ratio of 6:1.

on the Apion estate). For *P.Oxy.* XVI 1920.4–5, see pp. 120–21, 123 above.

other products: fish (*P.Oxy.* XVI 1923.9); meat (*P.Oxy.* 1923.10); *smēma* (cf. above, p. 134; *P.Oxy.* XVI 2051r, passim; *SB* XX 14625.31 [+ *BL* XII 227]); *stalagma* (cf. above, p. 134; *P.Oxy.* XVI 2051r, passim); vinegar (*P.Oxy.* XVI 1870.12, 2044, passim, XXVII 2480.31–32, 44, 46, 84, 97, 99, 119–22; *PSI* VIII 953.2, 4–6, 9, 12, 14, 37; *SB* XX 15181).

notes: In both *P.Oxy.* XIV 1720.5 and *P.Wash.Univ.* II 105.3–4, the *diploun* is the same size as the *sēkōma* (a decanting vessel; cf. below); in *P.Oxy.* XVI 1951 and in XVIII 2196r.11–12, it is the equivalent of the *knidion. Dipla* are described as "Oxyrhynchite" (cf. Kruit and Worp 2000b, 122–25) in *P.Oxy.* LXXII 4901.8. Both a *diploun mega* (probably in *P.Oxy.* LI 3631.6; cf. *SB* VIII 9683.19 [provenance unknown, IV]) and a *diploun mikron* (cf. *CPR* XIV 51.3 [HGv: provenance unknown, but see pp. 193–94; VI]) are attested. The jars in *P.Oxy.* XVI 1923 that are used for fish and meat are described as "sealed." The wine in *PSI* III 191 and 193 is described as *palaios,* but not consistently. The wine in *P.Oxy.* LXVIII 4683.3 is "Theban." See also *potaria* (below).

Casson (1939, 5–6) states that *diploun* simply meant "double jar" in the Roman period but was "unquestionably a measure" in the Byzantine period. With its variable capacities, this seems doubtful. *Diplous* modifies *knidion* (for which see below) in, for example, *P.Oxy.* LXIII 4375.2, and it seems plausible that *knidion* is often suppressed when *diploun* appears alone (cf. note to the text just cited). *Diplous* can modify, however, the names of other containers, for example, the *moustarion* (*SB* XVIII 13922.2) and the *keramion* (e.g., *Stud.Pal.* VIII 1177). Cf. also Kruit and Worp 1999, 117 n. 28.

διχώνιον (*dichōnion*): Cf. Kruit and Worp 2000a.

Oxyrhynchite attestations: *P.Laur.* IV 185.16 [where διχ( ) is resolved as δίχ(ωρα); see now *BL* XII 98; VII]; *P.Oxy.* LVIII 3942 (606); *SB* XXVI 16752.4, 18, 28 [where διχ( ) is resolved as δίχ(ωρα) or δίχ(οα); see now *BL* XII 247; 494–95].

capacities:—

other products:—

notes: *P.Stras.* V 394.5, v.1 (V–VI, Oxyrhynchus?), in which διχ( ) is resolved as δίχ(ωρον) or δίχ(ωρα), is probably another attestation; cf. now *BL* XII 261. [*P.IFAO* II 12b.10 is a late Oxyrhynchite text in which δι( ) has been read as δί(χωρα), but δι(πλᾶ) is no doubt correct; cf. *P.Wash.Univ.* II 98, 3n., and now *BL* XII 92.]

κεράμιον, κεραμίς (***keramion, -is***): The most common container for wine in earlier periods, it is rarely attested as such in the late Oxyrhynchite.

Oxyrhynchite attestations: *P.Oxy.* XIV 1776 (IV/V; cf. *BL* IX 188), XVI 1851 (VI–VII), 1924 (V–VI), 2043 (V), LIX 4002 (IV–V).

capacities: The late Oxyrhynchite texts provide no data; for earlier periods, cf. Kruit and Worp 1999, 118–19.

other products: cedar oil or a product containing this oil (*P.Oxy.* XVI 1924.5); cheese (*P.Oxy.* XIV 1776.11–12, LIX 4002.11); soap (*P.Oxy.* XVI 1924.4).

notes: Used for Rhodian wine in *P.Oxy.* XVI 1851.2–3.

κνίδιον (***knidion***): See Kruit and Worp 2000b, 104–10, and (with caution) Mayerson 2002 (which supersedes Mayerson 2000b).

Oxyrhynchite attestations: *O.Ashm.Shelt.* 51 (VI); *P.Flor.* I 65 (570–71); *P.Oxy.* I 140 (= *W.Chr.* 438; 550), 150 (= *Stud.Pal.* III 283; 590), 155 (VI), XVI 1896 (577), 1923 (V/VI), 1951 (485?; cf. Gonis 2004a, 199–200), 2012 (590), 2042 (V), 2047 (V), XVIII 2196r (after 587; see p. 16 n. 103 above), LVI 3862 (IV–V), LVIII 3960 (621); *PSI* VIII 954 (VI); *P.Stras.* VII 680? (see notes for *laē* below; VII); *P.Wash.Univ.* I 41 (V–VI); *P.Wisc.* I 11 (VII, after Islamic conquest; cf. *BL* VIII 201); *SB* I 1979 (VI), XXII 15368 (590).

capacities: 4 *sextarii* (*P.Wisc.* I 11.18–19); 8 *sextarii* (*P.Oxy.* LVIII 3960; cf. p. 99 n. 35 above; contra Mayerson 2002, 206–7.) The vessels in *O.Ashm.Shelt.* 51 are *tetraknidia* (*BL* XII 293).

other products: honey (*P.Oxy.* I 155.3, LVI 3862.23); olives (3862.23); veal (*P.Oxy.* XVI 1923.25; cf. *BL* IX 192); vinegar (3960.16, 22).

notes: referred to as *geouchika* (*P.Oxy.* I 140.21); referred to as *megala* (*P.Oxy.* LVI 3862.22–23); equated with *dipla* (*P.Oxy.* XVI 1951, XVIII 2196r.11–12).

κόλλαθον (***kollathon***): See Kruit and Worp 2000b, 136–40 and refs. Attested as a container for wine in, for example, *P.Ness.* III 95.39, but nothing in the text with the only late Oxyrhynchite attestation (*P.Oxy.* XVI 2052.7; 579) suggests such a use.

κοῦρι (***kouri***): This late vessel, commonly attested in connection with wine in the Arsinoite, is mentioned only in *P.Oxy.* XVI 1862 (ca. 624?; cf. *BL* VIII 250), where it is a container for wine (l. 7) and honey (l. 11c), and *SB* I 1960

(= *O.Petr.* 452, but the transcriptions differ; the *SB* transcription is more accurate), where it holds wine (l. 4, in *mikra kouri*), a pickled foodstuff (l. 5), and vinegar (l. 6; see Worp 2004, 570, for this problematic line). See further Kruit and Worp 1999, 110–11, but it should be noted that their discussion of the geographical distribution of this vessel is misleading.

κοῦφον (***kouphon***): Any empty jar; for discussion, see Mayerson 1997a. (The argument in *P.NYU* II 21, 2n., that there was "also a special type of jar named 'κοῦφον,'" is unconvincing.)

κρατήριον (***kratērion***): *P.Oxy.* XVI 2049 (VI) is the only secure late attestation of any provenance; in it, κρατήρια appear alongside διπλᾶ as wine measures. Possibly also in *P.Wash.Univ.* II 105.1 (VI–VII), where κρ( ) appears.

λάγυνος (***lagynos***): See Worp 2004, 561–64 and refs.

Oxyrhynchite attestations: *P.Berl.Zill.* 7 (574); *P.Oxy.* XVI 1862 (ca. 624?; *BL* VIII 250), 1923 (V/VI); *P.Stras.* V 394 (V–VI); *SB* XIV 11617 (580; cf. *BL* VIII 373), XX 14625 (V–VI).

capacities:—

other products: chickpeas (*SB* XX 14625.20 + Kruit and Worp 2000b, 99 n. 83 = *BL* XII 227); small fish (l. 14); grease (l. 21 + *BL* XI 230); radish oil (*P.Oxy.* XVI 1862.11b); salted fish (*P.Oxy.* XVI 1923.12 + Worp 2004, 562); dill/anise (l. 13).

notes: Referred to as *meston* in *SB* XIV 11617.16 (*BL* VIII 373) and *mikron* in *P.Stras.* V 394.5–6.

ⲗⲁⲏ/ⲗⲁⲩ (***laē, -u***): See Worp 2004, 564–65.

Oxyrhynchite attestations: *SB* I 1960 (VI/VII; = *O.Petr.* 452, but the transcription differs significantly; the *SB* transcription is more accurate), 1969 (VI), XX 14625 (V–VI), XXVI 16752 (494–95); *P.Stras.* V 394 (V–VI).

capacities: The (secure) Oxyrhynchite attestations provide no data. According to *P.Sarga*, pp. 23–24, the ⲗⲁⲏ was the same size as the *mikron knidion* and had a capacity of 5 *sextarii;* in *P.Stras.* VII 680, however, a ⲗⲁⲏ seems to be the equivalent of 12 knidia.

other products: honey (*SB* XX 14625.22; Italian honey in l. 26); mustard (l. 28); vinegar (l. 30).

notes: The adjective *megas* is used to describe the ⲗⲁⲏ in *P.Stras.* V 394.4, in *SB* I 1960 (but not in *O.Petr.* 452), and in the (unprovenanced) *CPR* XIV 51.3. Worp (2004, 565) suggests that this last piece and the other unprovenanced Greek papyrus, *P.Stras.* VII 680, come from the region around Oxyrhynchus; while there is other evidence to support this argument for the Vienna text, it should be noted that the Coptic attestations of the vessel are more geographically dispersed.

μικρόν (***mikron***): Appears (alone, substantively) as a wine vessel in *P.Wash.Univ.* II 105 (VI–VII). See 3n., and cf. *P.Sarga*, p. 24, where μεγάλα also appear.

μουστάριον ***(moustarion):***

Oxyrhynchite attestations: *P.Flor.* I 65 (570–71); *P.Oxy.* LXI 4132 (619).

capacities: The Oxyrhynchite attestations provide no data; see "notes."

other products:—

notes: All attestations are Byzantine. Both Oxyrhynchite texts are sales in advance, but the *moustaria* are among the extras (i.e., not the containers for the primary object of sale). In *CPR* VIII 63.2 (provenance unknown), *knidia moustaria* of five *sextarii* appear; in *SB* XVIII 13922.2 (ed.: unknown provenance),[13] a *diploun moustarion* is mentioned. Both Ricci (1924, 73–74) and Sijpesteijn (1985, 78) associate the *moustarion* with μοῦστος/μοῦσθος and the measure μοσθίον (*P.Oxy.* XII 1589.16).

ξέστης (***xestēs***): The Greek for *sextarius*. It is the base wine measure for military and related payments (cf., e.g., *P.Oxy.* XVI 1920 [563, cf. *BL* X 145], 2046 [563?, cf. p. 123 above], XXVII 2480.50–54, 57–60 [580/81, cf. pp. 92–95, 97 above]), and is also used for wine in most of the (official) commodity accounts published in *P.Oxy.* LI (3628–30, 3632–33; all V). A wine measure (in isolation) in *P.Oxy.* LXIII 4388 (423). For further discussion see Kruit and Worp 1999, 111–17, 119–20, which unfortunately does not address the assertion in Lang 1976, 76, that there were distinct *sextarii* for wine (.654 l) and oil (.728 l) after the third century. (Since a resolution for this metrological question is beyond the scope of the present work, I have used the frequently cited .546 l throughout.)[14]

ὀμφακηρά (***omphakēra***): See Woytek 1996, 237, and Kruit and Worp 2002a, 53–56.

---

13. "Unknown provenance": But the Michigan APIS web site notes "described as coming from Behneseh (Oxyrhynchus)."

14. Concerning the discussion in Schilbach 1970, 112–19, regarding the later *metron* standard, it should noted that *P.Grenf.* II 90 does not contain an early reference to the *metron monastēriakon*; the measure in question is simply that of the Monastery of Abba Agenes.

Oxyrhynchite attestations: *P.Oxy.* XVI 1870 (V), 1924 (V–VI), 2047 (V); *P.Wash.Univ.* II 105 (VI–VII); *SB* XX 14625 (V–VI).
capacities:—
other products: fish sauce (*P.Oxy.* XVI 1924.6), *propoma* (an apperitif; *P.Oxy.* XVI 2047.2).
notes: *P.Oxy.* XVI 1870, 12–13n.: "a rounded (grape-shaped) vessel is presumably meant." In *P.Wash.Univ.* II 105.3, 2 *sēkōmata* are decanted into 4 *omphakērai.* Note also *P.Oxy.* X 1343.6 (VI) with Kruit and Worp 2002a, 53–56 (though one should simply understand *kouphon* as "empty jar," not as a special vessel type).

ποτάρια (***potaria;*** **pl.):**
Oxyrhynchite attestations: Appears only in *P.Oxy.* XVI 2044 (probably 579). *BL* IX 192: < Latin *potorium,* "Trinkbecher."
capacities:—
other products:—
notes: Used in an account of vinegar. It stands in an uncertain grammatical relationship (apposition?) to *dipla* in lines 14, 16, and 19; unless the *oinocheiristēs* was only counting vessels (cf. p. 121 above), these *dipla* would seem to be of the same capacity as the others in the account.

σάμαθον (***samathon***): Occurs only in *P.Oxy.* X 1290 (V), where it is used for oil. (Cf. σάμβαθον, σάμφατον.) Cf. Mayerson 1998a; Kruit and Worp 2000b, 127–29; Kruit and Worp 2001, 87.

Σάμιον (***Samion***):
"Oxyrhynchite" attestations: *P.Oxy.* LVIII 3954 (611); *PSI* V 474 (VI). *P.Laur.* IV 185.16 (VII) has σάμοι, on which see Kruit and Worp 2000b, 129.
capacities:—
other products:—
notes: The cited texts are the only late attestations of any provenance. The first two of them explicitly concern the Apiones' *oikos* at Kynopolis; the third certainly comes from a "large estate" milieu (cf. *P.Oxy.* LXX 4794, 7n.) in the region of Oxyrhynchus, broadly construed. Not to be confused with the Samian ware described in Johnson and West 1949, 114–15.

σήκωμα (***sēkōma***): A measure, typically used for decanting; cf. Mayerson 1998b and 2001b.

Oxyrhynchite attestations: *P.Col.* VIII 245 (VI); *P.David* 10 (523); *P.Flor.* I 65 (570–71); *P.Harr.* I 100 (V); *P.Heid.* V 358 (524–25), 359 (VI–VII); *P.Laur.* IV 185 (VII); *P.Mich.* XI 608 (VI), XV 734 (572), 743 (622; cf. Gonis 2004a, 198); *P.Oxy.* XIV 1720 (VI), XVI 1896 (577), XLIX 3512 (492), LXI 4132 (619), LXXV 5069 (546); *P.Rein.* II 102 (V/VI; cf. Capron et al. 2004, 209); *PSI* VIII 881 (VI; note *BL* XII 253), X 1122 (VII; cf. Gonis 2005, 170–71); *P.Wash.Univ.* II 105 (VI–VII), inv. 60 (vol. III, in preparation); *SB* V 8264 (524; cf. *BL* VII 198 and XI 203), XXII 15725 (VI). (Note: After examining the digital image, I find the reading in *P.Col.* VIII 240.12 suspect.)

capacities: 4 *sextarii* (*P.Rein.* II 102.5; note *BL* IX 224); 5 *sextarii* (*P.Col.* VIII 245.9; *P.David* 10.4; *P.Heid.* V 358.3–4; *P.Mich.* XI 608.3–4, 6, XV 743.14–15; *P.Oxy.* LXI 4132.20–21, v.2; *P.Rein.* II 102.3–4; *PSI* VIII 881.5, X 1122.12–13, 36; *SB* XXII 15725.6); 6 *sextarii* (*P.Flor.* I 65.6–7, v.1; *SB* V 8264.12–13); 8 *sextarii* (*P.Heid.* V 359.2; *P.Mich.* XV 734.16–17; *P.Oxy.* XVI 1896.19–20, LXXV 5069, 14–16).[15]

other products:—

notes: All attestations are in texts dated to the fifth century or later. All but two texts are unequivocally from the Oxyrhynchite, and it seems likely that use was largely confined to that region—that is, the unprovenanced *P.Princ.* II 106 is probably an Oxyrhynchite text. The other text, *P.Laur.* III 84, is from the Small Oasis, but Oxyrhynchite influence would be expected. (See also Jördens 1992a, 231 n. 12, and Kruit and Worp 1999, 117). The *sēkōma* has the same capacity as the *diploun* (for which see above) in *P.Oxy.* XIV 1720 and *P.Wash.Univ.* II 105. They are described as *geouchika* in *P.Oxy.* XVI 1896, as "by the measure of St. Epimachos" in *P.Oxy.* LXI 4132 (probably; cf. 24–25n.), and as "of the son of Hellas" in *SB* XXII 15725.

σκεῦος (*skeuos*):

Oxyrhynchite attestations: *P.Oxy.* LVIII 3942 (606).

capacities: Rea (*P.Oxy.* LVIII 3942, 18–20n.) suggests that they are 8-*chous* jars, but use of the *chous* as a unit of capacity would be anachronistic. (Kruit and Worp 2000a is similarly skeptical.)

other products:—

notes: *P.Oxy.* XVI 1953 (419) is possibly relevant since an *oinopratēs* is involved.

15. Kruit and Worp (1999, 116; 2000b, 106) also indicate 7 *sextarii,* but they do not give an attestation.

σπαθίον (*spathion*): See Kruit and Worp 2001, refuting the argument in Mayerson 1998c that it was a wineskin.

Oxyrhynchite attestations: *P.Oxy.* I 155 (VI), XVI 2043 (V), XVI 2048 (V), LVI 3862 (IV–V), LXIII 4387 (401).

capacities: Perhaps 5–10 *sextarii* per *spathion*; cf. *P.Oxy.* LVI 3875, 2n. (This text is known by Kruit and Worp [2001], but it does not enter into their discussion of the vessel's capacity, which, on the basis of Epiphanius' *De mensuris et ponderibus,* they suggest was 12 *sext.*)

other products:—

notes: In *P.Oxy.* XVI 2048, *spathia* are containers for both wine from an Oasis (probably the Small Oasis; ll. 1–2) and local wine (l. 3).

σταμνίον (*stamnion*): Attested in classical literature as a vessel for wine (cf. Fleischer 1956, 40), it possibly appears in *P.Oxy.* XVI 1870.10–11 (V) as a container for radish oil.[16] Cf. also *O.Krok.* I 17, 4n.

χύμα (*chyma*): More common in the Arsinoite.

Oxyrhynchite attestations: *T.Varie* 8 (669).

capacities: 5 *sextarii* (l. 8)

other products:—

notes:—

16. Kruit and Worp (2000b, 136), however, suggest σ̣ι̣[κε|λικό]ν̣ for σ̣τ̣[α|μνίο]ν̣ (= *BL* XII 141).

Literature Cited

Abu-Lughod, Janet L.

1989 *Before European hegemony: The world system A.D. 1250–1350.* Oxford: Oxford University Press.

Abu-Lughod, Lila

1991 Writing against culture. In *Recapturing anthropology: Working in the present,* edited by Richard G. Fox, 137–62. Santa Fe: School of American Research Press.

Adams, Bertrand

1964 Paramoné *und verwandte Texte: Studien zum Dienstvertrag im Rechte der Papyri.* Neue Kölner rechtswissenschaftliche Abhandlungen 35. Berlin: W. de Gruyter.

Alston, Richard

2002 Managing the frontiers: Supplying the frontier troops in the sixth and seventh centuries. In *The Roman army and the economy,* edited by Paul Erdkamp, 398–419. Amsterdam: J. C. Gieben.

Amélineau, Emile, ed.

1895 *Monuments pour servir à l'histoire de l'Égypte chrétienne aux iv*[e]*, v*[e]*, vi*[e]*, et vii*[e] *siècles.* Mémoires publiés par les membres de la Mission archéologique française au Caire, vol. 4, fasc. 2. Paris: E. Leroux.

Anderson, Perry

1979 *Lineages of the absolutist state.* London: Verso.

Andreau, Jean

2002 Twenty years after Moses I. Finley's *The ancient economy.* In *The ancient economy,* edited by Walter Scheidel and Sitta von Reden, 33–49. Edinburgh: Edinburgh University Press.

Andreau, Jean, and Jérôme Maucourant

1999 À propos de la "rationalité économique" dans l'Antiquité gréco-romaine: Une interprétation des thèses de D. Rathbone. *Topoi* 9:47–102.

Ast, Rodney

2002 Papyri editae in memoriam Wm. Brashear I. *ZPE* 139:177–86.

Azzarello, Giuseppina

2006 *P.Oxy.* XVI 2039 e la nascita della *domus gloriosa* degli Apioni. *ZPE* 155: 207–28.

2007 Neue Papyruszeugnisse zur Apionenfamilie. In *Von Noricum nach Ägypten: Eine Reise durch die Welt der Antike,* edited by Karl Strobel and Renate Lafer, 251–61. Klagenfurt: Mohorjeva Hermagoras.
2010 Vecchi e nuovi personaggi della famiglia degli Apioni nei documenti papiracei. *PapCongr. XXV*:33–46.
Forthcoming *Il dossier della* domus divina *in Egitto.* Berlin: W. de Gruyter.

Bachrach, Bernard
1967 Was there feudalism in Byzantine Egypt? *JARCE* 6:163–66.

Bagnall, Roger S.
1977 Price in "sales on delivery." *GRBS* 18:85–96.
1993a *Egypt in late antiquity.* Princeton: Princeton University Press.
1993b Managing estates in Roman Egypt: A review article. *BASP* 30:127–35.
1993c Review of Rathbone 1991. *JRS* 83:254–56.
1995 *Reading papyri, writing ancient history.* Approaching the Ancient World. New York: Routledge.
1999 The date of *P.Kell.* I G. 62 and the meaning of χωρί ον. *CdE* 74:329–33.
2000 Vegetable seed oil is sesame oil. *CdE* 75:133–35.
2003 A vineyard in the Small Oasis: What was the dispute about? *JJP* 33:17–25.

Bagnall, Roger S., and Klaas A. Worp
1980 Grain land in the Oxyrhynchite nome. *ZPE* 37:263–64.

Bailey, Donald M.
1998 *Excavations at el-Ashmunein V: Pottery, lamps, and glass of the late Roman and early Arab periods.* British Museum Expedition to Middle Egypt. London: British Museum Publications.

Ballet, Pascale, and Maurice Picon
1987 Recherches préliminaires sur les origines de la céramique des Kellia (Égypte). *CCE* 1:17–48.

Banaji, Jairus
1992 Rural communities in the late empire. D.Phil. diss., Oxford University.
1997 Modernizing the historiography of rural labour: An unwritten agenda. In *Companion to historiography,* edited by Michael Bentley, 88–102. London: Routledge.
1998 Discounts, weight standards, and the exchange-rate between gold and copper: Insights into the monetary process of the sixth century. In *Atti dell'Accademia romanistica costantiniana: 12. convegno internazionale in onore di Manlio Sargenti,* edited by Giuliano Crifò and Stefano Giglio, 183–202. Naples: Edizioni scientifiche italiane.
1999 Agrarian history and the labour organisation of Byzantine large estates. In Bowman and Rogan 1999, 193–216.
2000 State and aristocracy from the late empire to Byzantium. In *Production and public powers in classical antiquity,* edited by Elio Lo Cascio and Dominic W. Rathbone, 92–99. Cambridge: Cambridge Philological Society.
2007 *Agrarian change in late antiquity: Gold, labour, and aristocratic dominance.* 2nd ed. Oxford: Oxford University Press.
2009 Aristocracies, peasantries, and the framing of the early Middle Ages. *Journal of Agrarian Change* 9:59–91.

Bang, Peter

2008 *The Roman bazaar: A comparative study of trade and markets in a tributary empire.* Cambridge Classical Studies. Cambridge: Cambridge University Press.

Beaucamp, Joëlle

1992 *Le statut de la femme à Byzance (4*[e]*–7*[e] *siecle).* Vol. 2, *Les pratiques sociales.* Travaux et mémoires du Centre de recherche d'histoire et civilisation de Byzance, monographies, no. 6. Paris: De Boccard.

2001 Apion et Praejecta: Hypothèses anciennes et nouvelles données. *REByz* 59:165–78.

2007 Byzantine Egypt and imperial law. In *Egypt in the Byzantine world, 450–700,* edited by Roger S. Bagnall, 271–87. Cambridge: Cambridge University Press.

Bell, Harold I.

1917 The Byzantine servile state in Egypt. *JEA* 4:86–106.

Benaissa, Amin

2006 An Oxyrhynchite sale on delivery from the reign of Mauricius. *Tyche* 21: 1–10.

2007 An estate overseer's work contract and the meaning of *exotikoi topoi. BASP* 44:75–86.

2009 *Rural settlements of the Oxyrhynchite nome: A papyrological survey.* Trismegistos Online Publications 4. Leuven: Trismegistos Publications. http://www.trismegistos.org/top.php.

Forthcoming The economic arrangements between the Apion estate and its *enapographoi georgoi.* In *Ricchezza e potere nell'Egitto bizantino: La famiglia degli Apioni ed altre casate ossirinchite,* edited by Giuseppina Azzarello. Berlin: W. de Gruyter.

Bergquist, Stig

1992 Considerations on yields, the distribution of crops, and the size of estates: Three Roman agricultural units. *C&M* 43:111–39.

Bloch, Marc

1966 *French rural history: An essay on its basic characteristics.* Translated by Janet Sondheimer. Berkeley and Los Angeles: University of California Press.

Bonneau, Danielle

1983 Communauté rurale en Égypte byzantine? *Recueils de la Société Jean Bodin* 41:505–23.

1993 *Le régime administratif de l'eau du Nil dans l'Égypte grecque, romaine et byzantine.* Probleme der Ägyptologie 8. Leiden: E. J. Brill.

Borsch, Stewart J.

2005 *The Black Death in Egypt and England: A comparative study.* Austin: University of Texas Press.

Bowman, Alan K.

1986 *Egypt after the pharaohs.* Berkeley: University of California Press.

Bowman, Alan K., and Eugene Rogan, eds.

1999 *Agriculture in Egypt from pharaonic to modern times.* Proceedings of the British Academy 96. Oxford: Oxford University Press for the British Academy.

Bowman, Alan K., et al., eds.
2007 *Oxyrhynchus: A city and its texts.* Graeco-Roman Memoirs 93. London: Egypt Exploration Society.
Breccia, Evaristo
1932 *Le musée gréco-romain, 1925–1931.* Bergamo: Istituto Italiano d'Arti Grafiche.
1933 *Le musée gréco-romain, 1932–1933.* Bergamo: Istituto Italiano d'Arti Grafiche.
Brenner, Robert
1985 Agrarian class structure and economic debate. In *The Brenner debate: Agrarian class structure and economic development in pre-industrial Europe,* edited by Trevor H. Ashton and Charles H. E. Philpin, 10–63. Cambridge: Cambridge University Press.
Brock, Sebastian
1981 The conversations with the Syrian Orthodox under Justinian (532). *OCP* 47:87–121.
Cameron, Averil
1979 A Nativity poem of the sixth century A.D. *CPh* 79:222–32.
1993 *The Mediterranean world in late antiquity.* Routledge History of the Ancient World. London: Routledge.
Capasso, Mario, et al., eds.
1990 *Miscellanea papyrologica in occasione del bicentenario dell'edizione della Charta Borgiana.* Pap.Flor. 19. Florence: Gonnelli.
Capron, Laurent, et al.
2004 Corrigenda aux P.Reinach II. *ZPE* 150:207–13.
Carrié, Jean-Michel
1997 "Colonato del Basso Impero": La resistenza del mito. In Lo Cascio 1997, 75–150.
2004 Vitalité de l'industrie textile à la fin de l'Antiquité: Considérations économiques et technologiques. *AnTard* 12:13–43.
Casson, Lionel
1939 Wine measures and prices in Byzantine Egypt. *TAPA* 70:1–16.
Chouliara-Raïos, Hélène, and Guy Wagner
1990 Στάγμα. *ZPE* 84:69–74.
Clarysse, Willy, and Katelijn Vandorpe
1997 Viticulture and wine consumption in the Arsinoite Nome (*P.Köln* V 221). *AncSoc* 28:67–73.
Cockle, Helen M.
1981 Pottery manufacture in Roman Egypt: A new papyrus. *JRS* 71:87–97.
Coleman, Tyler
2008 *Wine politics: How governments, environmentalists, mobsters, and critics influence the wines we drink.* Berkeley: University of California Press.
Comfort, Howard
1937 *Emphyteusis* among the papyri. *Aegyptus* 17:3–24.
Conning, Jonathan
2003 Latifundia economics. Working paper. Hunter College Department of Economics Working Papers 02/1, Hunter Collge: Department of Economics (PDF).

Constantinides, Elizabeth
1969 An Oxyrhynchus papyrus. *BASP* 6:55–58.
Crawford, John R.
1920 De Bruma et Brumalibus festis. *BZ* 23:365–95.
Crislip, Andrew T.
2005 *From monastery to hospital: Christian monasticism and the transformation of health care in late antiquity.* Ann Arbor: University of Michigan Press.
Dagron, Gilbert, et al.
2000 L'organisation et le déroulement des courses d'après le *Livre des cérémonies. T&MByz* 13:1–200.
Delbrück, Richard
1929 *Die Consulardiptychen und verwandte Denkmäler.* 2 vols. Studien zur spätantiken Kunstgeschichte 2. Berlin: W. de Gruyter.
Delmaire, Roland
1989 *Largesses sacrées et* res privata: *L'aerarium* *impérial et son administration du IV*e *au VI*e *siècle.* Collection de l'École française 121. Rome: École française de Rome.
1995 *Les institutions du Bas-Empire romain de Constantin à Justinien.* Vol. 1, *Les institutions civiles palatines.* Initiations au christianisme ancien. Paris: Éditions du Cerf.
Drew-Bear, Marie
1979 *Le nome hermopolite: Toponymes et sites.* Am.Stud.Pap. 21. Missoula: Scholars Press.
Dyer, Christopher
1998 Rural Europe. In *The new Cambridge medieval history,* vol. 7, edited by Christopher Allmand, 106–20. Cambridge: Cambridge University Press.
Dzierzbicka, Dorota
2005 Wineries and their elements in Graeco-Roman Egypt. *JJP* 35:9–91.
Egloff, Michel
1977 *Kellia: La poterie copte.* 2 vols. Recherches suisses d'archéologie copte 3. Geneva: Georg.
Empereur, Jean-Yves
1993 La production viticole dans l'Égypte ptolémaïque et romaine. In *La production du vin et de l'huile en Méditerranée,* edited by Jean-Pierre Brun and Marie-Claire Amouretti, 39–47. Athens: École française d'Athènes.
Erdkamp, Paul
2005 *The grain market in the Roman Empire: A social, political, and economic study.* Cambridge: Cambridge University Press.
Falivene, M. Rosaria
1997 *The Heracleopolite nome: A catalogue of the toponyms with introduction and commentary.* Am.Stud.Pap. 37. Atlanta: Scholars Press.
Fikhman, Isaak F.
1962 Remeslo i krupnoe imenie v vizantijskom Egipte po dannym greceskikh papirusov. *Palestinskij Sbornik* 7:51–88.
1968 On the structure of the Egyptian large estate in the sixth century. *PapCongr. XII*:127–32.

1973 Quelques considérations sur les données sociales et économiques des papyrus d'Oxyrhynchus d'époque byzantine. *JÖB* 22:15–21.
1974 Kurienland in Oxyrhynchos? Translated by Günter Poethke. In *Festschrift zum 150-jährigen Bestehen des Berliner Ägyptischen Museums,* 343–46. Berlin: Akademie-Verlag.
1980 Une caution byzantine pour des *coloni adscripticii.* In *Miscellanea papyrologica,* edited by Rosario Pintaudi, 67–77. Florence: Gonnelli.
1991 Esclaves et colons en Égypte byzantine. *AnPap* 3:7–17.
2006 *Wirtschaft und Gesellschaft im spätantiken Ägypten: Kleine Schriften Itzhak F. Fikhman.* Edited by Andrea Jördens. Historia Einzelschriften 192. Stuttgart: Franz Steiner Verlag.

Finley, Moses I.
1999 *The ancient economy.* Updated ed. Sather Classical Lectures 43. Berkeley: University of California Press.

Firestone, Ya'akov
1975 Crop-sharing economics in Mandatory Palestine—Part I. *Middle Eastern Studies* 11:3–23.

Fleischer, Rita M.
1956 Measures and containers in Greek and Roman Egypt. M.A. thesis, New York University.

Förster, Hans
1998 Neueditionen koptischer Texte aus der Wiener Papyrussammlung. *APF* 44:285–98.

Fournet, Jean-Luc
1998 Un nouvel épithalame de Dioscore d'Aphrodité adressé à un gouverneur civil de Thébaïde. *AnTard* 6:65–82.

Foxhall, Lin
1990 The dependent tenant: Land leasing and labour in Italy and Greece. *JRS* 80:97–114.

Foxhall, Lin, and Hamish A. Forbes
1982 *Sitometreia:* The role of grain as a staple food in classical antiquity. *Chiron* 12:41–90.

Franko, George F.
1988 *Sitometria* in the Zenon archive: Identifying Zenon's personal documents. *BASP* 25:13–98.

Gagos, Traianos, and Peter van Minnen
1992 Documenting the rural economy of Byzantine Egypt: Three papyri from Alabastrine. *JRA* 5:186–202.

Gagos, Traianos, and Pieter J. Sijpesteijn
1995 Settling a dispute in fourth-century Small Oasis (P. Mich. inv. no. 4008). *ZPE* 105:242–52.

Gascou, Jean
1972 La détention collégiale de l'autorité pagarchique dans l'Égypte byzantine. *Byzantion* 42:60–72.
1976a L'institution de bucellaires. *BIFAO* 76:143–56.
1976b Les institutions de l'hippodrome en Égypte byzantine. *BIFAO* 76:185–212.

1985 Les grands domaines, la cité et l'État en Égypte byzantine: Recherches d'histoire agraire, fiscale et administrative. *T&MByz* 9:1–90.
1989 La table budgétaire d'Antaeopolis (*P. Freer* 08.45 c–d). In *Hommes et richesses dans l'Empire byzantin*, edited by Jacques Lefort and Cécile Morrisson, 1:279–313. Paris: P. Lethielleux.
1998 Ducs, *praesides*, poètes et rhéteurs au Bas-Empire. *AnTard* 6:61–64.
2008 *Fiscalité et société en Égypte byzantine*. Bilans de recherche 4. Paris: Association des amis du Centre d'histoire et civilisation de Byzance.

Gascou, Jean, and Leslie MacCoull
1987 Le cadastre d'Aphroditô. *T&MByz* 10:103–58.

Gascou, Jean, and Klaas A. Worp
1988 *CPR* VII 26: Réédition. *Tyche* 3:103–10.
1990 Un dossier d'ostraca du VIe siècle: Les archives des huiliers d'Aphroditô. In Capasso et al. 1990, 217–44.

Goitein, Samuel D.
1983 *A Mediterranean society: The Jewish communities of the Arab world as portrayed in the documents of the Cairo Geniza*. Vol. 4, *Daily life*. Berkeley: University of California Press.

Gomaà, Farouk, et al.
1991 *Mittelägypten zwischen Samalūṭ und dem Gabal Abū Ṣīr: Beiträge zur historischen Topographie der pharaonischen Zeit*. Beihefte zum Tübinger Atlas des Vorderen Orients, Reihe B (Geisteswissenschaften), no. 69. Weisbaden: L. Reichert.

Gonis, Nikolaos
2000a A late Byzantine land lease from Oxyrhynchus: *P.Oxy.* XVI 1968 revised. *Tyche* 15:93–102.
2000b Six small Byzantine papyri from the Duke collection. *BASP* 37:71–81.
2002a Further texts from the Oxyrhynchus racing archive. *ZPE* 141:162–64.
2002b Studies on the aristocracy of late antique Oxyrhynchus. *Tyche* 17:85–97.
2004a Notes on Oxyrhynchus papyri III. *ZPE* 150:197–202.
2004b *P.Bingen* 135 and Flavius Apion I. *ZPE* 146:175–78.
2005 Seventh-century Oxyrhynchite documents in the Beinecke Library. *ZPE* 153:169–71.
2008 Notes on the aristocracy of Byzantine Fayum. *ZPE* 166:203–10.
2009 Prosopographica II. *APF* 55:90–95.

Grant, Mark, trans.
1997 *Dieting for an emperor: A translation of books 1 and 4 of Oribasius'* Medical compilations *with an introduction and commentary*. Studies in Ancient Medicine 15. Leiden: E. J. Brill.

Grenfell, Bernard P.
1896/97 Oxyrhynchus and its papyri. *Archaeological Report* 6:1–12.

Grenfell, Bernard P., and Arthur S. Hunt
1902/3 Excavations at Hîbeh, Cynopolis, and Oxyrhynchus. *Archaeological Report* 12:1–9.
1903/4 Excavations at Oxyrhynchus. *Archaeological Report* 13:14–17.
1904/5 Excavations at Oxyrhynchus. *Archaeological Report* 14:13–17.

1905/6 Excavations at Oxyrhynchus. *Archaeological Report* 15:8–16.
1906/7 Excavations at Oxyrhynchus. *Archaeological Report* 16:8–11.

Grover, B. R.
1964 Nature of *dehat-l-taaluqa* (*zamindari* villages) and the evolution of the *taaluqdari* system during the Mughal age. *Indian Economic Social History Review* 2.3:259–88.

Hagedorn, Dieter, and Bärbel Kramer
2004 Fünf neue Papyri des *comes* Johannes (P. Hamb. inv. 532, 533, 538, 547 und P. Heid. inv. 1800 + 1843) und Neuabdruck von *P.Harris* I 91. *APF* 50:158–71.

Hagedorn, Dieter, and Fritz Mitthof
2004 *P.Paramone* 18: Ein neu plaziertes Fragment oder: *Cave restauratorem chartarum. ZPE* 149:157–58.

Haldon, John F.
1984 *Byzantine praetorians: An administrative, institutional, and social survey of the Opsikion and Tagmata, c. 580–900.* Poikila Byzantina 3. Bonn: R. Habelt.

Hallagan, William S.
1978 Self-selection by contractual choice and the theory of sharecropping. *Bell Journal of Economics* 9:344–54.

Hardy, Edward R.
1931 *The large estates of Byzantine Egypt.* Studies in History, Economics, and Public Law 354. New York: Columbia University Press.

Harrauer, Hermann, and Pieter. J. Sijpesteijn
1986 *P. Princ.* II 96 und Schreibübungen. *ZPE* 64:115–16.

Hengstl, Joachim
1972 *Private Arbeitsverhältnisse freier Personen in den hellenistischen Papyri bis Diokletian.* Bonn: R. Habelt.

Herrmann, Johannes
1958 *Studien zur Bodenpacht im Recht der graeco-aegyptischen Papyri.* Münch.Beitr. 41. Munich: C. H. Beck.

Hickey, T. M.
1998 Notes on some Cairo papyri from Byzantine Oxyrhynchus. *ZPE* 123:161–64.
1999 Notes on some papyri from the Byzantine Oxyrhynchite. *Tyche* 14:325–27.
2001 A "public house" but closed: "Fiscal participation" and economic decision making on the Oxyrhynchite estate of the Flavii Apiones. Ph.D. diss., University of Chicago.
2007 Aristocratic landholding and the economy of Byzantine Egypt. In *Egypt in the Byzantine world, 450–700*, edited by Roger S. Bagnall, 288–308. Cambridge: Cambridge University Press.
2008 An inconvenient truth? *P.Oxy.* 18.2196 verso, the Apion estate, and *fiscalité* in the late antique Oxyrhynchite. *BASP* 45:87–100.
2009 Writing histories from the papyri. In *Oxford Handbook of Papyrology*, edited by Roger S. Bagnall, 495–520. Oxford: Oxford University Press.
Forthcoming Collecting for the *despotēs:* The account of Stephanos *pronoētēs* (*P.Oxy.* VI 999 *descr.*). In *Ricchezza e potere nell'Egitto bizantino: La famiglia degli Apioni ed altre casate ossirinchite*, edited by Giuseppina Azzarello. Berlin: W. de Gruyter.

Hoffman, Philip T.
1984 The economic theory of sharecropping in early modern France. *Journal of Economic History* 44:309–20.

Husson, Geneviève
1983 Oikia: *Le vocabulaire de la maison privée en Égypte d'après les papyrus grecs.* Publ.Sorb.Pap. 2. Paris: Publications de la Sorbonne.

Jackson, Roland S.
2008 *Wine science: Principles and applications.* 3rd ed. Burlington: Academic Press.

Jakab, Éva
1999 Guarantee and jars in sales of wine on delivery. *JJP* 29:33–44.
2001 "Wo gärt der verkaufte Wein?" Zur Deutung der Weinlieferungskäufe in den graeco-ägyptischen Papyri. In *Symposion 1997,* edited by Eva Cantarella and Gerhard Thür, 295–318. Cologne: Verlag Hermann Böhlaus Nachfolger.

Johnson, Allan C., and Louis C. West
1949 *Byzantine Egypt: Economic studies.* Princ.Stud.Pap. 6. Princeton: Princeton University Press.

Jones, Arnold H. M.
1958 The Roman colonate. *Past and Present* 13:1–13.
1964 *The later Roman Empire, 284–602: A social, economic, and administrative survey.* Norman: University of Oklahoma Press.

Jördens, Andrea
1986a Die ägyptischen *Symmachoi. ZPE* 66:105–18.
1986b Teilpachtverträge aus dem Arsinoites. *ZPE* 65:107–22.
1992a Fünf neue Symmachos-Papyri. *ZPE* 92:219–231.
1992b Μίσθωσις τῶν ἔργων: Ein neuer Vertragstyp. *PapCongr. XIX.2:*259–70.
1993 Kaufpreisstundungen. *ZPE* 98:263–82.
1997 P.Bodl. I 82: Ein προχρεία-Vertrag. *ZPE* 116:81–84.
1999 Die Agrarverhältnisse im spätantiken Ägypten. *Laverna* 10:114–52.
2001 Papyri und private Archive: Ein Diskussionsbeitrag zur papyrologischen Terminologie. In *Symposion 1997,* edited by Eva Cantarella and Gerhard Thür, 253–68. Cologne: Verlag Hermann Böhlaus Nachfolger.

Kark, Ruth, and David Grossman
2003 The communal (musha') village of the Middle East and North Africa. In *Policies and strategies in marginal regions: Summary and evaluations,* edited by Walter Leimgruber et al., 223–36. Aldershot: Ashgate.

Keenan, James G.
1978 The case of Flavia Christodote: Observations on *PSI* I 76. *ZPE* 29:191–209.
1981 On village and *polis* in Byzantine Egypt. *PapCongr. XVI:*479–85.
1993 Papyrology and Byzantine historiography. *BASP* 30:137–44.
2007 Byzantine Egyptian villages. In *Egypt in the Byzantine world, 450–700,* edited by Roger S. Bagnall, 226–43. Cambridge: Cambridge University Press.

Kehoe, Dennis P.
1992 *Management and investment on estates in Roman Egypt during the early empire.* Pap.Texte Abh. 40. Bonn: R. Habelt.
1997 *Investment, profit, and tenancy: The jurists and the Roman agrarian economy.* Ann Arbor: University of Michigan Press.

2007 *Law and the rural economy in the Roman Empire.* Ann Arbor: University of Michigan Press.

2008 Economics and the law of water rights in the Roman Empire. In *Vers une gestion intégrée de l'eau dans l'Empire romain,* edited by Ella Hermon, 243–52. Rome: "L'Erma" di Bretschneider.

Kloppenborg, John S.

2006 *The tenants in the vineyard.* Wissenschaftliche Untersuchungen zum Neuen Testament 195. Tübingen: Mohr Siebeck.

Kramer, Johannes

1990 Die Bedeutung von σπανέλαιον. *ZPE* 81:261–64.

1992 *Roga. ZPE* 94:185–90.

Krüger, Julian

1990 *Oxyrhynchos in der Kaiserzeit: Studien zur Topographie und Literaturrezeption.* Europäische Hochschulschriften, Reihe III (Geschichte und ihre Hilfswissenschaften), no. 441. Frankfurt am Main: Peter Lang.

Kruit, Nico

1992a Local customs in the formulas of sales of wine for future delivery (A supplement to *P.Heid.* V). *ZPE* 94:167–84.

1992b The meaning of various words related to wine: Some new interpretations. *ZPE* 90:265–76.

Kruit, Nico, and Klaas A. Worp

1999 Metrological notes on measures and containers of liquids in Graeco-Roman and Byzantine Egypt. *APF* 45:96–127.

2000a ΔΙΧΟΝΙΟΝ = "two-chous jar." *Mnemosyne* 53:345–46.

2000b Geographical jar names: Towards a multi-disciplinary approach. *APF* 46: 65–146.

2001 The *spathion* jar in the papyri. *BASP* 38:79–87.

2002a A seventh-century list of jars from Edfu. *BASP* 39:47–56.

2002b Two notes on Byzantine containers. *MBAH* 21.2:44–52.

Kula, Witold

1976 *An economic theory of the feudal system: Towards a model of the Polish economy, 1500–1800.* Translated by Lawrence Garner. Foundations of History Library. London: NLB.

Laiou, Angeliki E.

1977 *Peasant society in the late Byzantine Empire: A social and demographic study.* Princeton: Princeton University Press.

Lang, Mabel L.

1976 *Graffiti and dipinti.* Athenian Agora 21. Princeton: American School of Classical Studies at Athens.

Laniado, Avshalom

2002 *Recherches sur les notables municipaux dans l'Empire protobyzantin.* Travaux et Mémoires monographies 13. Paris: Association des amis du Centre d'histoire et civilisation de Byzance.

Le Roy Laudurie, Emmanuel

1985 A reply to Robert Brenner. In *The Brenner debate: Agrarian class structure and economic development in pre-industrial Europe,* edited by Trevor H. Ashton and Charles H. E. Philpin, 101–6. Cambridge: Cambridge University Press.

Liebeschuetz, John H. G. W.

2001 *The decline and fall of the Roman city.* Oxford: Oxford University Press.

Lo Cascio, Elio, ed.

1997 *Terre, proprietari e contadini dell'impero romano: Dall'affitto agrario al colonato tardoantico.* Rome: Nuova Italia Scientifica.

Love, John R.

1991 *Antiquity and capitalism: Max Weber and the sociological foundations of Roman civilization.* London: Routledge.

Luzzatto, Maria J.

1996 P. Vat. gr. 52: Trasporto di vino dall'Egitto a Costantinopoli nel VII sec. d. C. *ZPE* 114:153–56.

MacCoull, Leslie S. B.

1990 Φιλοτιμία in Byzantine papyrus documents. *Tyche* 5:63–66.

Manning, Joseph G., and Ian Morris, eds.

2005 *The ancient economy: Evidence and models.* Stanford: Stanford University Press.

Marcone, Arnaldo

1997 Il colonato tardoantico: I problemi aperti. In Lo Cascio 1997, 225–39.

Maresch, Klaus

1994 *Nomisma und Nomismatia: Beiträge zur Geldgeschichte Ägyptens im 6. Jahrhundert n. Chr.* Abhandlungen der Nordrhein-Westfälischen Akademie der Wissenschaften, Sonderreihe Papyrologica Coloniensia. Opladen: Westdeutscher Verlag.

Martin, Alain

1994 Archives privées et cachettes documentaires. *PapCongr. XX*:569–78.

Mayerson, Philip

1992 The Gaza "wine" jar (Gazition) and the "lost" Ashkelon jar (Askalōnion). *Israel Exploration Journal* 42:76–80.

1997a A note on κοῦφα "empties." *BASP* 34:47–52.

1997b The role of flax in Roman and Fatimid Egypt. *JNES* 56:201–8.

1998a Epiphanius' *sabitha* in Egypt: Σάμβαθον/σάμφαθον/σάμαθον. *BASP* 35:215–18.

1998b ΣΗΚΩΜΑΤΑ—"Standard" measures for decanting wine. *BASP* 35:153–58.

1998c Σπάτιον/σπάθιον—A wineskin. *ZPE* 121:226–28.

2000 The *knidion* jar in Egypt: Popular, made in Egypt, and of unknown capacity. *ZPE* 131:165–67.

2001a Jar stoppers and the sealing of winejars. *ZPE* 136:217–20.

2001b Σηκώματα: Measures of wine, not jars. *BASP* 38:97–101.

2002 The enigmatic *knidion:* A wine measure in late Roman/Byzantine Egypt? *ZPE* 141:205–9.

2005 Σῖτος/πυρός in Egypt as deliberate mixtures of wheat with barley. *BASP* 42:51–62.

2006 *Mega kankellon* and *metron* in late receipt and expenditure accounts. *BASP* 43:107–11.

Mazza, Roberta
1995 Ricerche sul pagarca nell'Egitto tardoantico e bizantino. *Aegyptus* 75:169–242.
1997 Φλ. Ἀπίων γενόμενος πρωτοπατρίκιος. *Simblos* 2:211–19.
1998 *P.Oxy.* XVI 1911 e i conti annuali dei *pronoetai. ZPE* 122:161–72.
2001 *L'archivio degli Apioni: Terra, lavoro e proprietà senatoria nell'Egitto tardoantico.* Munera 17. Bari: Edipuglia.
2004 Noterelle prosopografiche in margine ad alcune pubblicazioni recenti riguardanti gli Apioni. *Simblos* 4:263–80.
2005 Dalla *Bruma* ai *Brumalia:* Modelli di cristianizzazione tra Roma e Costantinopoli. In *Diritto romano e identità cristiana,* edited by Alessandro Saggioro, 161–71. Rome: Carocci.
2007 Gli affitti fondiari dell'Ossirinchite nell'età tardoantica. *PapCongr. XXIII:* 439–46.
2008 Review of Sarris 2006. *Journal of Agrarian Change* 8:150–56.
McCabe, Anne E.
2007 *A Byzantine encyclopaedia of horse medicine: The sources, compilation, and transmission of the* Hippiatrica. Oxford Studies in Byzantium. Oxford: Oxford University Press.
Metcalf, David M.
2004 *Byzantine lead seals from Cyprus.* Texts and Studies of the History of Cyprus 47. Nicosia: Cyprus Research Center.
Meyer, Elizabeth A.
2004 *Legitimacy and law in the Roman world:* Tabulae *in Roman belief and practice.* Cambridge: Cambridge University Press.
Mickwitz, Gunnar
1937 Economic rationalism in Graeco-Roman agriculture. *EHR* 52:577–89.
Minnen, Peter van
1986 The volume of the Oxyrhynchite textile trade. *MBAH* 5:88–95.
2004 Greek papyri and Coptic Studies, 1996–2000. *CoptCongr. VII:*423–35.
Mitthof, Fritz
2001 Annona militaris: *Die Heeresversorgung im spätantiken Ägypten. Ein Beitrag zur Verwaltungs- und Heeresgeschichte des römischen Reiches im 3. bis 6. Jh. n.Chr.* Pap.Flor. 32. Florence: Gonnelli.
2008 Das Dioskoros-Archiv und die militärischen Reformen Justinians in der Thebais. In *Les archives de Dioscore d'Aphrodité cent ans après leur découverte,* edited by Jean-Luc Fournet, 247–59. Paris: De Boccard.
Montserrat, Dominic
2007 News reports: The excavations and their journalistic coverage. In Bowman et al. 2007, 28–37.
Morelli, Federico
1995 Sulle retribuzioni nell'Egitto bizantino: Il caso dei ποταμῖται. *PapCongr. XXI:*727–37.
1996 *Olio e retribuzioni nell'Egitto tardo: 5.–8. d.C.* Florence: Istituto Papirologico "G. Vitelli."
1997 Τιμή e μισθός: Vendita e prestazione di lavoro: Osservazioni sulle relazioni economiche tra artigiani e proprietà nell'Egitto bizantino. *Comunicazioni* (Istituto Papirologico "G. Vitelli") 2:7–29.

1998 P. Vindob. G 42920 e la φιλοτιμία di ʿUmar b. Marwān. *ZPE* 121:219–21.
2004 Il λαχανόσπερμον, il ῥαφανέλαιον, e il sesamo: Olii e oleaginose fantasma. *ZPE* 149:138–42.
2008 Zwischen Poesie und Geschichte: Die "*Flagornerie*" des Dioskoros und der dreifache Dux Athanasios. In *Les archives de Dioscore d'Aphrodité cent ans après leur découverte,* edited by Jean-Luc Fournet, 223–45. Paris: De Boccard.

Mossner, Ernest C., and Ian S. Ross, eds.
1987 *The correspondence of Adam Smith.* 2nd ed. Oxford: Oxford University Press.

Nagel, Peter, trans.
1983 *Das Triadon: Ein sahidisches Lehrgedicht des 14. Jahrhunderts.* Wissenschaftliche Beiträge der Martin-Luther-Universität Halle-Wittenberg, 1983/23. Halle an der Saale: Martin-Luther-Universität Halle-Wittenberg.

Nour ed-Din, S.
1898 Conditions des fellahs en Égypte. *Revue de l'Islam* 3:5–7.

Oleson, John Peter
1984 *Greek and Roman mechanical water-lifting devices: The history of a technology.* Phoenix Supplementary Volumes 16. Toronto: University of Toronto Press.

Palme, Bernhard
1994 Flavius Sarapodorus, ein *agens in rebus* aus Hermopolis. *APF* 40:43–68.
1998a Flavius Epiphanius, *comes domesticorum. Eirene* 34:98–116.
1998b Flavius Strategius Paneuphemos und die Apionen. *ZSS r.a.* 115:289–322.
2007 The imperial presence: Government and army. In *Egypt in the Byzantine world, 450–700,* edited by Roger S. Bagnall, 244–70. Cambridge: Cambridge University Press.
2008 Flavius Flavianus—von Herakleopolis nach Konstantinopel? *BASP* 45: 143–69.

Papaconstantinou, Arietta
1996 La liturgie stationnale à Oxyrhynchos dans la première moitié du 6e siècle: Réédition et commentaire du *P.Oxy.* XI 1357. *REByz* 54:135–59.
2001 *Le culte des saints en Égypte des Byzantins aux Abbassides: L'apport des inscriptions et des papyrus grecs et coptes.* Le monde byzantine. Paris: CNRS Éditions.

Papathomas, Amphilochios
1999 Μοναστήριον ἐποικίου Σκυταλίτιδος: Ein neues Kloster im Oxyrhynchites. *ZPE* 128:167–68.

Peacock, David P. S., and David P. Williams
1986 *Amphorae and the Roman economy: An introductory guide.* Longman Archaeology Series. New York: Longman.

Perpillou-Thomas, Françoise
1993a *Fêtes d'Égypte ptolémaïque et romaine d'après la documentation papyrologique grecque.* Studia Hellenistica 31. Leuven: Universitas Catholica Lovaniensis.
1993b Les *Brumalia* d'Apion II. *Tyche* 8:107–9.

Petrie, W. M. Flinders, et al.
1925 *Tombs of the courtiers and Oxyrhynkhos.* Publications of the British School of Archaeology in Egypt and the Egyptian Research Account, no. 37. London: British School of Archaeology in Egypt and the Egyptian Research Account.

Pintaudi, Rosario
2007 The Italian excavations. Translated by Livia Capponi. In Bowman et al. 2007, 104–8.
Poethke, Günter
1969 Epimerismos: *Betrachtungen zur Zwangspacht in Ägypten während der Prinzipatszeit.* Pap.Brux. 8. Brussels: Fondation égyptologique Reine Élisabeth.
Pruneti, Paola
1981 *I centri abitati dell'Ossirinchite: Repertorio toponomastico.* Pap.Flor. 9. Florence: Gonnelli.
Pudney, Stephen, et al.
1998 An econometric model of farm tenures in fifteenth-century Florence. *Economica* 65:535–56.
Purcell, Nicholas
1985 Wine and wealth in ancient Italy. *JRS* 75:1–19.
Rāghib, Yūsuf
1991 La plus ancienne lettre arabe de marchand. In *Documents de l'Islam medieval: Nouvelles perspectives de recherche,* edited by Yūsuf Rāghib, 1–9. Cairo: Institut français d'archéologie orientale.
Rapp, Claudia
1997 Ritual brotherhood in Byzantium. *Traditio* 52:285–325.
Rathbone, Dominic W.
1990 Villages, land, and population in Graeco-Roman Egypt. *PCPhS,* n.s., 36: 103–42.
1991 *Economic rationalism and rural society in third-century A.D. Egypt: The Heroninos archive and the Appianus estate.* Cambridge Classical Studies. Cambridge: Cambridge University Press.
1997 Prices and price-formation in Roman Egypt. In *Économie antique: Prix et formation des prix dans les économies antiques,* edited by Jean Andreau et al., 183–224. Saint-Bertrand-de-Comminges: Musée archéologique départemental.
2001 The "Muziris" papyrus (*SB* XVIII 13167): Financing Roman trade with India. In *Alexandrian studies II in honour of Mostafa El Abbadi,* edited by Mohammed Abd-el-Ghani et al., 39–50. Alexandria: Archaeological Society of Alexandria.
Rea, John
1963 Additions and corrections to documentary papyri in Ox. Pap. Vol. XXVII. *JEA* 49:180–81.
Reil, Theodor
1913 *Beiträge zur Kenntnis des Gewerbes im hellenistischen Ägypten.* Borna bei Leipzig: Robert Noske.
Rémondon, Roger
1955 L'Édit XIII de Justinien, a-t-il promulgué en 539? *CdE* 30:112–21.
1961 Soldats de Byzance d'après un papyrus trouvé à Edfou. *Recherches de Papyrologie* 1:41–93.
1964 *La crise de l'Empire romain, de Marc Aurèle à Anastase.* Nouvelle Clio: L'histoire et ses problèmes 11. Paris: Presses Universitaires de France.

1966 Papyrologica: *PSI* V 472; *P.Bodl.* d. 54; *P.Lond.* III 982; *P.Herm. Rees* 79; *P.Goth.* 9; *P.Vars.* 31. *CdE* 41:161–79.

Ricci, Clotilde

1924 *La coltura della vite e la fabbricazione del vino nell'Egitto Greco-romano.* Studi della Scuola papirologica, vol. 4, pt. 1. Milan: R. Accademia scientifico-letteraria.

Richards, Alan

1978 Land and labor on Egyptian cotton farms, 1882–1940. *Agricultural History* 52:503–18.

Richter, Tonio S.

2005 What's in a story: Cultural narratology and Coptic child donation documents. *JJP* 35:237–64.

Rostovtzeff, Michael I.

1930 The decay of the ancient world and its economic explanations. *Economic History Review* 2:197–214.

1957 *The social and economic history of the Roman Empire.* 2nd ed., revised by Peter M. Fraser. Oxford: Oxford University Press.

Rouillard, Germaine

1928 *L'administration civile de l'Égypte byzantine.* 2nd ed. Paris: P. Geuthner.

Rowlandson, Jane

1994 Crop rotation and rent payment in Oxyrhynchite land leases: Social and economic interpretations. *PapCongr. XX*:495–99.

1996 *Landowners and tenants in Roman Egypt: The social relations of agriculture in the Oxyrhynchite nome.* Oxford Classical Monographs. Oxford: Clarendon.

2005 The organisation of public land in Roman Egypt. *CRIPEL* 25:173–96.

Ruffing, Kai

1999 *Weinbau im römischen Ägypten.* Pharos: Studien zur griechisch-römischen Antike 12. St. Katharinen: Scripta Mercaturae.

Ruffini, Giovanni

2008 *Social networks in Byzantine Egypt.* Cambridge: Cambridge University Press.

2010 Review of Sarris 2006. *Topoi* 16:627–40.

Russo, Simona

1999 *I gioielli nei papiri di età greco-romana.* Florence: Istituto Papirologico "G. Vitelli."

Saller, Richard P.

2005 Framing the debate over growth in the ancient economy. In *The ancient economy: Evidence and models,* edited by Joseph G. Manning and Ian Morris, 223–38. Stanford: Stanford University Press.

Samuel, Alan E.

1965 The role of *paramone* clauses in ancient documents. *JJP* 15:221–311.

Sarris, Peter

2004a The origins of the manorial economy: New insights from late antiquity. *EHR* 119:279–311.

2004b Rehabilitating the great estate: Aristocratic property and economic growth in the late antique East. In *Recent research on the late antique countryside,* edited by William Bowden et al., 55–71. Leiden: E. J. Brill.

2006 *Economy and society in the age of Justinian.* Cambridge: Cambridge University Press.

2009 Social relations and the land: The early period. In *A social history of Byzantium,* edited by John Haldon, 92–111. Blackwell: Malden.

Satyanarayana, Adapa

1985 The zamindari crisis in Andhra, 1900–39. *Studies in History* 1:111–36.

Schaffner, Julie A.

1995 Attached farm labor, limited horizons, and servility. *Journal of Development Economics* 47:241–70.

Scheidel, Walter

2000 Slaves of the soil. *JRA* 13:727–32.

2010 Real wages in early economies: Evidence for living standards from 1800 BCE to 1300 CE. *JESHO* 53:425–62.

Scheidel, Walter, and Sitta von Reden, eds.

2002 The ancient economy. Edinburgh: Edinburgh University Press.

Schilbach, Erich

1970 *Byzantinische Metrologie.* Handbuch der Altertumswissenschaft 12.4. Munich: C. H. Beck.

Schmitt, Oliver

1994 Die *Buccellarii:* Eine Studie zum militärischen Gefolgschaftswesen in der Spätantike. *Tyche* 9:147–74.

Schnebel, Michael

1925 *Die Landwirtschaft im hellenistischen Ägypten.* Münch.Beitr. 7. Munich: C. H. Beck.

Shelton, John C.

1990 New texts from the Oxyrhynchus racing archive. *ZPE* 81:265–66.

Sijpesteijn, Petra M.

2007 The Arab conquest of Egypt and the beginning of Muslim rule. In *Egypt in the Byzantine world, 450–700,* edited by Roger S. Bagnall, 437–59. Cambridge: Cambridge University Press.

Sijpesteijn, Pieter J.

1982 The measure τὸ διδιπλοῦν. *ZPE* 48:124.

1985 Small Byzantine texts. *ZPE* 61:75–84.

1992 The meanings of ἤτοι in the papyri. *ZPE* 90:241–50.

1994 P.Mich. inv. 3425: A correction. *ZPE* 103:97.

Sirks, A. J. Boudewijn

1993 Reconsidering the Roman colonate. *ZSS r.a.* 110:331–69.

2008 The colonate in Justinian's reign. *JRS* 98:120–43.

Stathakopoulos, Dionysios C.

2004 *Famine and pestilence in the late Roman and early Byzantine Empire: A systematic survey of subsistence crises and epidemics.* Birmingham Byzantine and Ottoman Monographs 9. Aldershot: Ashgate.

Syrkou, Angeliki

1995 L'agricoltura nell'Egitto tardo-romano e bizantino secondo i papiri. In *Atti del V Seminario internazionale di Papirologia,* edited by Mario Capasso, 127–34. Lecce: Congedo.

Talbot, Alice-Mary

1990 The Byzantine family and the monastery. *DOP* 44:119–29.

Tchernia, André

1986 *Le vin de l'Italie romaine: Essai d'histoire économique d'après les amphores.* Bibliothèque des Écoles françaises d'Athènes et de Rome 261. Rome: École française de Rome.

Thurman, William S.

1964 The thirteen edicts of Justinian translated and annotated. Ph.D. diss., University of Texas.

Tomber, Roberta, and David Williams

2000 Egyptian amphorae in Britain and the western provinces. *Britannia* 31: 41–54.

Torallas Tovar, Sofia

2001 Los *riparii* en los papiros del Egipto tardoantiguo. *Aquila legionis* 1:123–51.

2006 Violence in the process of arrest and imprisonment in late antique Egypt. In *Violence in late antiquity: Perceptions and practices,* edited by Harold A. Drake, 103–12. Aldershot: Ashgate.

Udovitch, Abraham L.

1999 International trade and the medieval Egyptian countryside. In Bowman and Rogan 1999, 267–85.

Van Alfen, Peter G.

1996 New light on the 7th-c. Yassi Ada shipwreck: Capacities and standard sizes of LRA1 amphoras. *JRA* 9:189–213.

Vera, Domenico

1992–93 Schiavitù rurale e colonato nell'Italia imperiale. *Scienze dell'antichità: Storia, archeologia, antropologia* 6-7:291–339.

Verhoogt, Arthur M. F. W.

1998 *Menches,* komogrammateus *of Kerkeosiris: The doings and dealings of a village scribe in the late Ptolemaic period (120-110 B.C.).* Pap.Lugd.Bat. 29. Leiden: E. J. Brill.

Vogt, Christine, et al.

2002 Notes on some of the Abbasid amphorae of Istabl ʿAntar-Fustat (Egypt). *BASOR* 326:65–80.

Volbach, Wolfgang F.

1976 *Elfenbeinarbeiten der Spätantike und des frühen Mittelalters.* 3rd ed. Kataloge vor- und frühgeschichtlicher Altertümer 7. Mainz am Rhein: Von Zabern.

Whittington, Graeme

1973 Field systems of Scotland. In *Studies of field systems in the British Isles,* edited by Alan R. H. Baker and Robin A. Butlin, 530–79. Cambridge: Cambridge University Press.

Wickham, Christopher J.

1988 Marx, Sherlock Holmes, and late Roman commerce. *JRS* 78:183–93.

2005 *Framing the early Middle Ages: Europe and the Mediterranean, 400–800.* New York: Oxford University Press.

Wipszycka, Ewa

1965 *L'industrie textile dans l'Égypte romaine.* Komitet Nauk o Kulturze Antycznej Polskiej Akademii Nauk, Archiwum filologiczne 9. Wrocław: Zakład Narodowy im. Ossolińskich.

1972 *Les ressources et les activités économiques des églises en Égypte du IVᵉ au VIIIᵉ siècle.* Pap.Brux. 10. Brussels: Fondation égyptologique Reine Élisabeth.

Worp, Klaas A.
2000 A survey of ἁπλᾶ, δι(δι)πλᾶ, and τριπλᾶ measures in the papyri. *ZPE* 131: 145–49.
2004 Notes on Coptic containers of liquids in Greek papyri. *CoptCongr. VII:* 553–72.
2005 Two further chronological notes on Byzantine documents. *ZPE* 151:153–58.
Woytek, Bernhard
1996 Idem Hermeias? Das Fragment eines Rechnungsbuches aus dem spätantiken Ägypten. *Tyche* 11:229–42.
Ziegler, Renate
1995 Bemerkungen zu verschiedenen Urkunden. *ZPE* 106:189–94.
Zuckerman, Constantine
1988 "Legio V Macedonica" in Egypt: *CPL* 199 revisited. *Tyche* 3:279–287.
1995 The hapless recruit Psois and the mighty anchorite, Apa John. *BASP* 32: 183–94.
1998 Two reforms of the 370s: Recruiting soldiers and senators in the divided empire. *REByz* 56:79–139.
2004 *Du village à l'Empire: Autour du registre fiscal d'Aphroditô (525/526).* Centre de recherche d'histoire et civilisation de Byzance. Monographies 16. Paris: Association des amis du Centre d'histoire et civilisation de Byzance.

Index Locorum

An asterisk following a locator indicates that the text is corrected or discussed in substance in this location. *t* indicates that the reference appears in a table.

B. Other Texts

Subject Index